Eye to the Sky

Exploring Our Atmosphere *2nd Ed.*

STEVEN BUSINGER

"*The wonder of the world, the beauty and the power,
the shapes of things, their colours, lights, and shades;
these I saw. Look ye also while life lasts.*"

On an old English tombstone

Steven Businger
Department of Atmospheric Sciences
School of Ocean, Earth Science and Technology
University of Hawaii

"Eye to the Sky – Exploring Our Atmosphere, Second Edition," by Steven Businger. ISBN 978-1-62137-888-4 (softcover).

Published 2016 by Virtualbookworm.com Publishing Inc., P.O. Box 9949, College Station, TX 77845, US.

PREFACE

The atmosphere, which envelops and sustains us, is continually changing. It is possible to walk outside at any moment and observe quite a show, sometimes of great beauty and sometimes of great violence. The primary goal of this book is to promote an awareness of our atmospheric environment. This goal is approached in two ways: through the art of making careful observations in the context of the scientific method, and through the art of data analysis and image interpretation.

Progress in science was revolutionized during the renaissance through development of the scientific method, in which careful observation leads to a prediction or hypothesis, followed by experimentation and finally the formulation of a theory. The following example of this combination of observation and reasoning applied in the scientific method is a classic experiment in atmospheric science. In 1643 Torricelli had invented the first barometer by immersing a glass tube in a dish of mercury and showing that a column of mercury would rise up the tube so as just to balance the weight of the air above the dish, thus providing the first measure of atmospheric pressure. Four years later, Pascal made one of the first completely scientific predictions. He reasoned that since atmospheric pressure is due to the weight of the air above you, there should be less pressure when you climb a mountain, and so the mercury in the barometer should not rise as high. Upon climbing the Puy du Dome in the French Alps with the help of friends, Pascal confirmed his prediction. News of his successful prediction caused great celebration all across Europe, because it represented a triumph of the new scientific method.

A captivating world of weather is unveiled to those who approach the experiments and exercises in this book, with a spirit of curiosity like that of Pascal. The experiments in this book are designed to give students of diverse academic backgrounds an opportunity to explore and understand firsthand the underlying physical principles of our everyday atmospheric environment through the art of making and interpreting observations. The materials for most of the experiments described can be obtained inexpensively from local stores. The book also includes a variety of exercises that rely on analysis of provided weather observations. These activities are designed to foster understanding of the wide range of observations made of our atmospheric environment, with an emphasis on pattern recognition and anticipating the future state of the atmosphere. Many of the labs and analysis exercises presented here can be enhanced with widely available web resources, many of which are identified in the labs (see also the Appendix: Online Resources).

Shorter experiments can be combined to take up an allotted laboratory period. Other activities, such as the analysis exercises in Chapter 6, will take longer and a subset can be assigned. Not all the activities provided in this book can be covered in a single college semester. The range of activities allows instructors to make a selection that suits the emphasis of their accompanying lecture course.

ACKNOWLEDGMENTS

Thanks are due to lab instructors at North Carolina State University and the University of Hawaii for field-testing the exercises in this manual and anonymous reviewers for their suggestions for improvements. The author is grateful to Nancy Hulbirt and Mary McVicker for assistance with figure drafting and to Diane Henderson for editing of the manuscript. Thanks to Brooks Bays for his design of the book cover.

ABOUT THE AUTHOR

Steven Businger is a professor of meteorology at the University of Hawaii. He received his PhD in Atmospheric Sciences at the University of Washington where he studied the evolution of arctic storms. He has since authored more than eighty articles on a range of subjects in the field of atmospheric science including winter storms, severe thunderstorms, hurricanes, pollution dispersion, and acid rain.

TABLE OF CONTENTS

Chapter 1 Origin and Composition of the Atmosphere _____ *1*

Lab 1: Eye to the Sky _____ 8

Lab 2: Oxygen in the Atmosphere _____ 13

Lab 3: Particles in the Air _____ 17

Lab 4: Introduction to Analyzing a Weather Map _____ 20

Lab 5: Indirect Measurements _____ 27

Chapter 2 Pressure And Temperature _____ *32*

Lab 6: Atmospheric Pressure _____ 37

Lab 7: Crushing Cans _____ 39

Lab 8: Exploding Popcorn _____ 42

Lab 9: Density, Temperature, and Pressure _____ 47

Chapter 3 Sunlight and the Earth's Climate _____ *50*

Lab 10: Even You Emit Radiation _____ 55

Lab 11: Distance from the Sun _____ 58

Lab 12: Surface Heating _____ 61

Lab 13: Solar Radiation _____ 66

Lab 14: Local Climate _____ 70

Chapter 4 Weather's Invisible Fuel _____ *74*

Lab 15: Measuring Moisture in the Air _____ 80

Lab 16: Wet-Bulb Temperature _____ 84

Lab 17: Evaporation and Surface Area _____ 88

Lab 18: The Effect of Salt on Vapor Pressure _____ 91

Chapter 5 Clouds And Precipitation _____ *94*

Lab 19: The Water Cycle _____ 103

Lab 20: Recycled Water-The Hydrologic Cycle _____ 106

Lab 21: Formation of Clouds _____ 109

Lab 22: Rain Makers _____ 112

Lab 23: Archimedes' Eureka Moment _____ 116

Lab 24: Convection _____ 120

Lab 25: Weather Balloons and Radiosondes _____ 127

Chapter 6 Wind, the General Circulation, and Storms _____ *136*

Lab 26: Wind: Air in Motion _____ 146

Lab 27: The Coriolis Effect _____ 148

Lab 28: Forces, Wind, and Weather Maps _____ 154

Lab 29: Jet Stream Winds _____ 160

Lab 30: Analyzing a Winter Storm _____ 166

Lab 31: Blizzards _____ 176

Chapter 7 Weather Forecasting _____ *180*

Lab 32: Weather Radar's X-Ray Vision _____ 193

Lab 33: Doppler Radar can See the Wind _____ 197

Lab 34: Satellites – Our Eyes in Space _____ 202

Lab 35: Satellite Analysis of a Winter Storm _____ 206

Lab 36: Numerical Weather Prediction _____ 210

Lab 37: Weather Forecasting _____ 214

Chapter 8 Severe Thunderstorms and Hurricanes _____ *218*

Lab 38: Investigating Hail _____ 226

Lab 39: Voices in the Heavens _____ 230

Lab 40: Storm Chasing _____ 234

Lab 41: Tracking Hurricane Inikl _____ 241

Lab 42: Flash-Flood Forecasting _____ 248

Lab 43: Hurricanes and Flooding _____ 252

Lab 44: Wind Generated Waves _____ 258

Chapter 9 Atmospheric Pollution and Global Warming _____ 272

Lab 45: Exhausting Problems _____ 277

Lab 46: Understanding Acidity _____ 280

Lab 47: Monitoring Acid Rain _____ 284

Lab 48: The Atmosphere Effect and Global Warming _____ 290

Lab 49: Volcanic Eruption _____ 297

Chapter 10 Sky Lights _____ 300

Lab 50: Bending Light _____ 306

Lab 51: Creating Rainbows _____ 311

Lab 52: Investigating Refraction _____ 314

Lab 53: Aerosols, Visibility, and the Color of the Sky _____ 317

Appendix: Online Resources _____ 321

Photograph Credits _____ 322

Chapter 1 Origin and Composition of the Atmosphere

"You cannot depend on your eyes if your imagination is out of focus."
– Mark Twain

The life-sustaining part of our atmosphere is precariously thin; roughly half of the mass of the atmosphere is below ~ 3 miles. As travelers know, commercial jets all have oxygen masks for their passengers, should the cabin suddenly loose pressure to the outside at cruising altitudes (±7 miles). If the Earth were likened to an onion, the thickness of the entire atmosphere would be akin the thickness of the onion's skin.

The Earth's present atmosphere is a mixture of gases, mostly nitrogen and oxygen, and large numbers of suspended particles. The concentrations of the most abundant components of the gaseous mixture, called air, are uniform through the lowest 80 km of the atmosphere and are essentially constant with time. On the other hand, concentrations of certain important gases such as water vapor and ozone vary substantially from place to place and from time to time. Carbon dioxide is fairly uniformly distributed, but its concentration has been increasing continually since the beginning of the industrial revolution in the nineteenth century. Although not a toxic gas, carbon dioxide is important in the radiation balance of the Earth and its increase has been linked to a gradual warming observed in the global climate.

The atmosphere contains huge numbers of particles from natural sources, such as the oceans and volcanoes, and from human inventions, such as motor vehicles, power stations, and industrial plants. A combination of air and the suspended particles is called an *aerosol*. Some particles, liquid and solid, play important roles in cloud, rain, and snow formation. Other particles are hazardous to health and costly to society.

The force or pressure exerted by the atmosphere is simply the weight of the air in a column (of unit cross section) extending to the top of the atmosphere. *Atmospheric pressure* is a maximum at the Earth's surface and always diminishes with increasing height because the mass of air above you decreases as you move to higher altitude. At sea level, atmospheric pressure averages about 1,013 millibars (~29.9 inches of mercury). In English units, this amounts to ~14.7 pounds per square inch. At an altitude of 50 km, the pressure is about 0.85 mb. Contours, or lines of constant pressure shown on weather maps, are called *isobars*.

It is well known that, on average, air temperature decreases with height through the lower atmosphere. This behavior is consistent with the observation that the atmosphere is reasonably transparent to sunlight, and thus the sun-warmed Earth substantially heats the air from below. This layer, where temperature decreases with height, is known as the *troposphere*. It ends at a level called the tropopause and is surmounted by the *stratosphere*, through which the temperature is constant or increases with height. Observations show several other higher layers of decreasing and increasing temperature (See Fig. 1.1).

In the stratosphere, chemical reactions, stimulated in part by the absorption of ultraviolet radiation from the sun, lead to the establishment and maintenance of a layer of ozone. The absorption of ultraviolet radiation also accounts for the warmth of the stratosphere. This layer extends from about 10 to 50 km and is extremely important to life on Earth. The ozone greatly reduces the amount of ultraviolet radiation reaching the ground. Ultraviolet rays can cause skin cancer and affect other biological processes. Various substances, such as chlorofluorocarbon gases, introduced into the atmosphere by human activities, pose a threat to the ozone layer.

In the upper layers of the atmosphere, air densities and pressures are very low; gaseous atoms and molecules exist in relatively small concentrations. The absorption of ultraviolet radiation causes electrons to be stripped from some gaseous species. This process accounts for a deep region of charged particles known as the *ionosphere*. Some long-distance radio communications systems still depend on reflections from the ionosphere. On some occasions, the ionosphere is disturbed, and radio transmissions via the ionosphere become ineffective.

At the uppermost reaches of the atmosphere, high-speed particles from the sun are guided towards polar regions by the Earth's magnetic field. They collide with air molecules and cause electrons to be freed. When recombination occurs and the air molecules return to their original states, light is emitted. The resulting brilliant displays of light are called the *aurora borealis* in the Northern Hemisphere, the *aurora australis* in the Southern Hemisphere. The more common names for these two phenomena are the northern and southern lights, respectively.

Origin of the Earth's Atmosphere

The Earth's atmosphere originated from an out-gassing of dissolved gases from inside the Earth as the planet cooled and the crust solidified. A clue to the composition of the early atmosphere can be obtained by analyzing the nature of the gases escaping from present volcanoes; these include nitrogen - N_2, water vapor - H_2O, sulfur dioxide - SO_2, carbon dioxide - CO_2, and various other trace gases such as argon. One gas that is very important in the Earth's present atmosphere but conspicuously absent in volcanic emissions is oxygen - O_2.

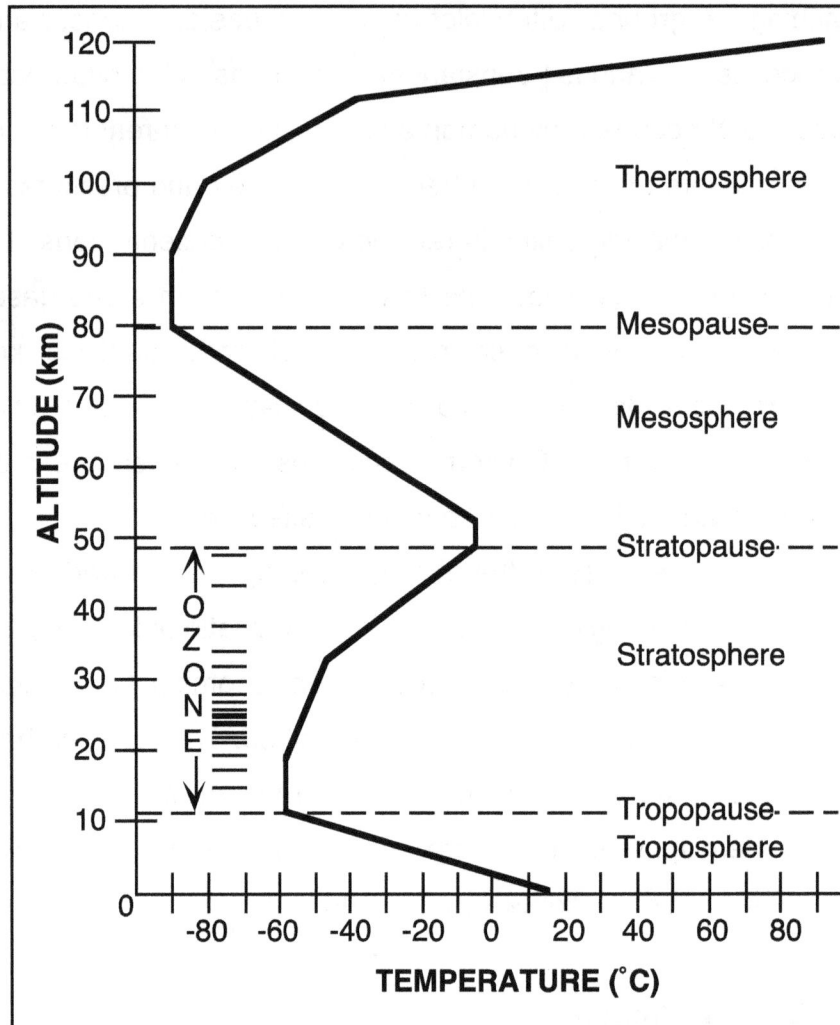

Figure 1.1 Temperature structure of the Earth's atmosphere

The question arises: Was the early Earth's atmosphere devoid of oxygen? Since oxygen is chemically active (it's an oxidizer that causes rust, for example), analysis of rocks that were exposed to the Earth's atmosphere very early in its evolution can give a clue as to the presence or absence of oxygen at that time. The fact that these ancient rocks show no evidence of oxidation suggests that the Earth's earliest atmosphere actually was devoid of oxygen.

How did oxygen get added to the Earth's atmosphere? Scientist have hypothesized that very intense solar radiation in the primitive atmosphere split the oxygen from the hydrogen in water. Another theory, which is supported by strong evidence, suggests that the main source of oxygen in the Earth's atmosphere came from plant life. In the process called photosynthesis, plant life consumes carbon dioxide and emits oxygen. Once oxygen appeared, ozone (O_3) also appeared. In the geologic record there is evidence for a very rapid expansion of the plant life over the Earth's surface once oxygen and ozone appeared in the atmosphere. This suggests that the addition of oxygen and ozone to the atmosphere occurred very quickly following the emergence of plant life in the oceans.

Measures and Conversions

This book is about observing the atmosphere. A visible satellite photograph of a storm system may be worth a thousand words (or more), but to understand the atmosphere more fully, we need to make measurements, like wind speed, and express the information as a number. The trouble with observations is that the magnitude of the number depends on the unit used. For example, we know that 40 mph is not the same as 40 km/hr, which is not the same as 40 m/s. To reduce the confusion associated with data collection, scientists have come to agree on a standard system of units called the International System (SI), which has its roots in the metric system. The basic units for the SI system are, meter (length), second (time), kilogram (weight), and degree Kelvin (temperature). In this book a variety of units (metric, English, SI) are used in a conscious effort to familiarize students with these units and their conversions. For convenience, Table 1.1 provides common unit conversions.

SI allows the sizes of units to be made bigger or smaller by the use of prefixes (Table 1.2). For example, the electrical unit of a watt is not a big unit even in terms of ordinary household use, so it is referred to as 1000 watts at a time. The prefix for 1000 is *kilo* so we use kilowatts (kW) as our unit of measurement. For makers of electricity, or bigger users such as industry, it is common to use megawatts (MW) or even gigawatts (GW).

Table 1.1 Common Unit Conversions

Length	1 kilometer (km) = 1000 (m) = 3281 feet (ft) = 0.62 miles (mi)
	1 mi = 5280 ft = 1.61 km = 0.87 nautical mile (nm)
	1 centimeter (cm) = 0.39 inch (in)
	1 in = 2.54 cm
	1 m = 1.093613 yards = 3.28084 ft
Time	1 day = 24 hour (hr) = 60
Mass	1 kilogram (kg) = 2.2 pounds (lb)
Speed	1 nautical mile per hour = 1 knot (kt) = 1.15 miles per hour (mph) =
	1.85 kilometers per hour (km/hr) = 0.514 meters per second (m/s)
	1 mph = 1.61 kilometers per hour (km/hr) = 0.45 meters per second (m/s)

The symbol used for *micro* is the Greek letter µ pronounced "miew." Nearly all of the SI prefixes are multiples (kilo to giga) or sub-multiples (milli to nano) of 000. However, these are inconvenient for many purposes and so *hecto, deca, deci,* and *centi* are also used. *Deca* also appears as *deka* (*da*) or (*dk*) in the USA and continental Europe. So much for standards! Table 1.2 provides a range of prefixes with their symbols.

Table 1.2 Orders of magnitude and scientific notation

giga	(G)	1 000 000 000	*(a thousand millions = a billion)*	$=10^9$
mega	(M)	1 000 000	*(a million)*	$=10^6$
kilo	(k)	1 000	*(a thousand)*	$=10^3$
hecto	(h)	100	*(a hundred)*	$=10^2$
deca	(da)	10	*(ten)*	$=10^1$ 1
deci	(d)	0.1	*(a tenth)*	$=10^{-1}$
centi	(c)	0.01	*(a hundredth)*	$=10^{-2}$
milli	(m)	0.001	*(a thousandth)*	$=10^{-3}$
micro	(µ)	0.000 001	*(a millionth)*	$=10^{-6}$
nano	(n)	0.000 000 001	*(a billionth)*	$=10^{-9}$

6

The atmosphere contains phenomena that encompass a large range of time and space scales, from tiny aerosol and cloud droplets to winter storms and planetary-scale jet streams. To deal with very large or very small numbers, meteorologists use *scientific notation,* convenient shorthand that makes unruly numbers manageable (Table 1.2). To convert from regular notation to scientific notation, the exponent is a count of how many decimal places the decimal point must move to the left (or right) of the decimal point in the number 1. So the number for one million = 1,000,000. = 1. x 10^6 = 10^6, because the decimal place has moved six places to the left. Similarly, a negative exponent means that the decimal place has been moved to the right; e.g., 10^{-6} = 0.000,001.

Lab 1: Eye to the Sky

INTRODUCTION

Have you ever driven down a road or hiked a trail when the spectacle of the sky was so striking that you stopped dead in your tracks?...if only you had a camera handy.   Even nature's less remarkable skies often include relatively rare and photogenic events.   Here are some tips to becoming a good sky observer and capturing those stunning images.

Key number one: The most important ingredient in becoming a successful observer of the sky is making time to be outside.   Keep an observant eye.   Make a habit of searching the sky whenever you exit your home, office or any building.   Many visually striking weather phenomena in the atmosphere are fleeting and often occupy only a fraction of the heavens.   Therefore, persistent attention to details of the sky will result in dramatic images.   Sometimes the brightness of the sun and our healthy tendency to avoid looking towards it causes us to miss some of the most beautiful displays of atmospheric optics (e.g., halos, iridescence or coronas) produced by a variety of thin clouds close to the sun.   In looking for such phenomena cover the sun with your extended hand and use of a good pair of sunglasses to enhance the contrast in the clouds, and to protect your eyes from the sun's harmful ultraviolet rays.

Key number two: keep a camera with you.   In recent years this has become easier with the arrival of lightweight digital and film cameras that combine increased capabilities with ease of use.   Many professional photographers prefer single-lens reflex cameras, in which metering and composing are achieved through the lens.   However, publishable photographs can be obtained with modern point-and-click models.   It can be awkward carrying a camera with you everywhere you go, but as soon as you decide to leave it behind the best rainbow or most dramatic sunset you have ever seen will appear (a corollary of Murphy's law).   I keep a camera in my car at most times and still I'm caught off guard.   It is not enough to have a camera close at hand; you also need film (memory card) and live batteries, which tend to run out just as the funnel cloud starts to form.   One solution is to keep plenty of film (extra memory) and extra batteries on hand.   It's best to keep extra film stored in a cool dry place such as a refrigerator. If you are interested in selling your work, it is useful to note that many markets prefer slides, but increasingly digital imagery is accepted. Images need to have sufficient resolution for the printed size (e.g., a 6-mb TIFF image will result in a sharp 8.5x11 print. Slower film speeds (≤100 ASA) give finer grain resolution and seem to provide the best results.

Key number 3: Careful composition is the third key to good sky photography. Try to balance the areas of the photograph. To make your compositions more balanced and compelling, divide the viewfinder into thirds in your mind. Then place the subject or focus of your photograph at one of four resulting points of intersection.

Choose the foreground of your sky photos carefully. For example, reflections of the sky in a lake or pond can give added interest to the foreground and create dynamic images. Similarly, the use of tree branches to frame a special cloud can add interest to an area of an image that otherwise lacks detail. On the other hand avoid clutter, such as telephone wires, that distract from your subject. Choose the right lens to show the subject to best advantage. Telephoto lenses are great for isolating areas of special interest in the sky. A strong telephoto is needed to capture a good example of a mirage or the fleeting "green flash" as the sun sets over water. Similarly, the best lens for catching halos and rainbows is a wide angle. To photograph an entire double rainbow requires a 90°-viewing angle. Ironically, wide-angle lenses are more challenging to compose with than telephoto lenses for the very reason that their field of view is so wide.

Some hints for taking special photos: Despite its fleeting nature, good photographs of lightning are surprisingly simple to obtain provided you approach the subject with caution and a bit of patience. On the next stormy night, choose a protected place (e.g., inside a truck) and place your camera on a sturdy tripod (or sand bag) to steady it against jarring from the wind. Select the shutter speed setting that allows the lens diaphragm to remain open while the shutter release is depressed (use the "b" setting for the shutter speed selection in cameras with manual override). Then press the shutter release and wait. Once a bright flash of lightning is observed through the viewfinder release the shutter. The lightning will expose the film much as

9

a flash bulb would. To get a really "striking" image leave the lens open for the duration of several lightning strokes.

When shooting halos and coronas that surround the sun you must point the camera directly at the sun. Contrary to intuition, shooting into the sun does not harm your camera, but will often result in underexposure of the subject you're trying to capture. To avoid this problem, it is advantageous to block out the sun with an object such as a treetop or flagpole, and if your camera allows, spot meter the subject for correct exposure. When shooting near the sun, shade the front of the lens to eliminate reflections that are internal to the lens, resulting in those artificial hexagon-shaped flares in the photos.

Some special cautions: Despite the opportunity to acquire stunning photographs, weather phenomena that inspire great awe such as tornadoes and hurricanes must be approached with extreme caution, and are perhaps best left to professional meteorologists. Hurricanes in particular make poor subjects to approach, not only in the threat that they represent to your health, but additionally your presence can interfere with critical emergency evacuation, rescue and relief efforts.

Parting shot: On your next excursion pack a camera, and keep your eyes to the sky. You will be rewarded for your persistence and patience in more ways than just good photographs. Happy hunting.

ACTIVITY

OBJECTIVE: The purpose of this lab is to promote an awareness of our natural environment through observation.

It may require some patience to observe many of the interesting phenomena that occur in our environment. Therefore, this lab is placed early in this book to give students ample time to observe. Not all phenomena are as spectacular as a rainbow or a thunderstorm; however, variations in the shades of the sky or the evolution of a cloud provide interesting subjects. Once you've learned to look for them you may well be surprised at how common some optics such as halos or sundogs are.

Many of the potential subjects for this lab are discussed later in this book. For example, a section on atmospheric optics appears at the end of the book, in keeping with the order of this topic's treatment in most textbooks. Therefore, students are encouraged to peek ahead to future sections for ideas and explanations for this assignment. Some instructors may wish to postpone the due date for this lab to the latter half of the semester or repeat the assignment (for extra credit?).

MATERIALS:
§ pen & paper
§ digital camera, film camera, or miscellaneous artistic supplies
§ sunglasses for observing optical phenomena near the sun
§ patience and creativity

Make a point of personally observing atmospheric, geophysical, or atmospheric optical phenomena. Any observation that you personally witness during the current semester, and has a relation to the subject matter of this course, is fair game. Creativity in the choice of your observation and in doing the assignment are encouraged

Some examples include:
Atmospheric optics--rainbows, halos, color variations in the sky, etc
Cloud formation and motions, condensation of water on a glass
Dispersion of pollution--cigarettes, smokestacks
Behavior of waves breaking, stream flow, effects of wind on water
Climatic effects on landforms
Doppler effects--passing cars, trains, etc.

These are just a few suggestions and you are definitely not limited to the above list. The more specific your choice of observation, the easier it will be to fulfill each of the four steps in the procedure section below. The finished assignment should be one or two type written pages, plus an illustration.

PROCEDURE:

1. Describe the setting of your observation, including the date, time, place and any circumstances that contribute to the phenomenon. (*Note: this observation must be made during the current semester, however, it need not necessarily take place outdoors*)

2. Describe the phenomenon in detail. Pay careful attention to detail. Use of your creative writing abilities is encouraged.

3. Illustrate (photograph or artwork, etc.) the phenomenon or a relevant aspect of the setting or physical mechanism(s) involved. Try to preserve the relative scales in the artwork. Creativity in any artwork will be appreciated. Photography is encouraged.

4. Give a concise explanation of what physically is producing the phenomenon.

Lab 2: Oxygen in the Atmosphere

INTRODUCTION

Earth's atmosphere is composed of a mixture of gases. However, excluding water vapor, which can vary from 0 to ~4%, ~99% of the atmosphere is made up of just two gases—oxygen and nitrogen. Because many people have heard so much about oxygen and because we know we must breathe it in order to live, it is often thought that the atmosphere is all or mostly oxygen. Another way that the presence of oxygen is evident is the oxidation process. Oxygen has the ability to rust metals, make apples turn brown, and allows a candle to burn. The oxidation process, in fact, changes a substance and often creates a new one altogether. The chemical reaction that causes rusting uses up oxygen from the air. In the activity below, if you are patient, you will see the effects of oxygen in the air being used up in a chemical reaction and be able to gauge how much of the air around us is made-up of oxygen. Near the ocean metal objects are observed to rust more quickly than elsewhere. In part this is due to higher humidity near the beach, but in part it is the result of salt in the air. How does the addition of salt affect the results of this experiment?

Sometimes a similar experiment is done in which a lit candle is used in the place of steel wool. The candle experiment has the advantage that it can be done very quickly. However, when a carbon-based fuel is burned in the presence of sufficient oxygen, carbon dioxide is produced and the volume of the gas produced will equal that of the gas consumed. Changes in volume of the air observed in this case are the result of the expansion and contraction of air as it is heated by the candled and cooled subsequently (see Lab 7: Crushing Cans).

ACTIVITY

OBJECTIVE: The goal of this experiment is to observe the affect of oxidation and gauge the percent of oxygen in the atmosphere.

MATERIALS:

§ 2 tall narrow jars with labels removed (8-oz jars)

§ steel wool (untreated, no soap)

§ 1 clear plastic container at least 2 in (5 cm) high and large enough to hold both jars

§ two rubber bands

§ ruler

§ water

§ cellophane

§ marking pen that can write on glass jar

PROCEDURE:

1. Wet a piece of steel wool and push it into the bottom of one jar so that the steel wool stays when the jar is turned upside down. If it falls down, use a larger piece of steel wool.

2. Pour water into the clear plastic container until the water is about 1.5 in (3-4 cm) deep. Turn both jars upside down and place each straight down into the water next to each other. In some of the trials salt can be added to the water in the container.

3. Tilt the jars so that air can escape and then stand the jars back up in the water so that the height of the water inside the jars is the same as the height of the water in the container.

4. Place a rubber band around each jar right at the water level in the jar. This will mark the water level in case it changes. Let the jars sit undisturbed for about two days.

5 Place some cellophane across the open part of the plastic container to reduce evaporation.

5. After one week note the water level in the jar with the steel wool (Jar 1) compared to the water level in the jar with no steel wool (Jar 2). Carefully mark the two levels with a marking pen on Jar 2 (without steel wool).

6. Fill Jar 2 with water up to the mark you made corresponding to Jar 1's level. Measure the volume with the graduated cylinder and record it in the Data Table below.

14

7. Refill Jar 2 with water up to the level corresponding to Jar 2's level. In doing so you are measuring the volume of air in Jar 2. Subtract the second volume you measured from the first and record it in the Data Table.

8. To calculate the percentage of oxygen in the air, use this equation:

   Percentage of O_2 in air = $\dfrac{(Jar\ 1's\ Volume - Jar\ 2's\ Volume)}{Jar\ 2's\ Volume} \times 100$

9. Record the data for all the other groups taking part in this activity in the Data Table.

10. From your multiple trials, calculate an average percentage of oxygen in the air.

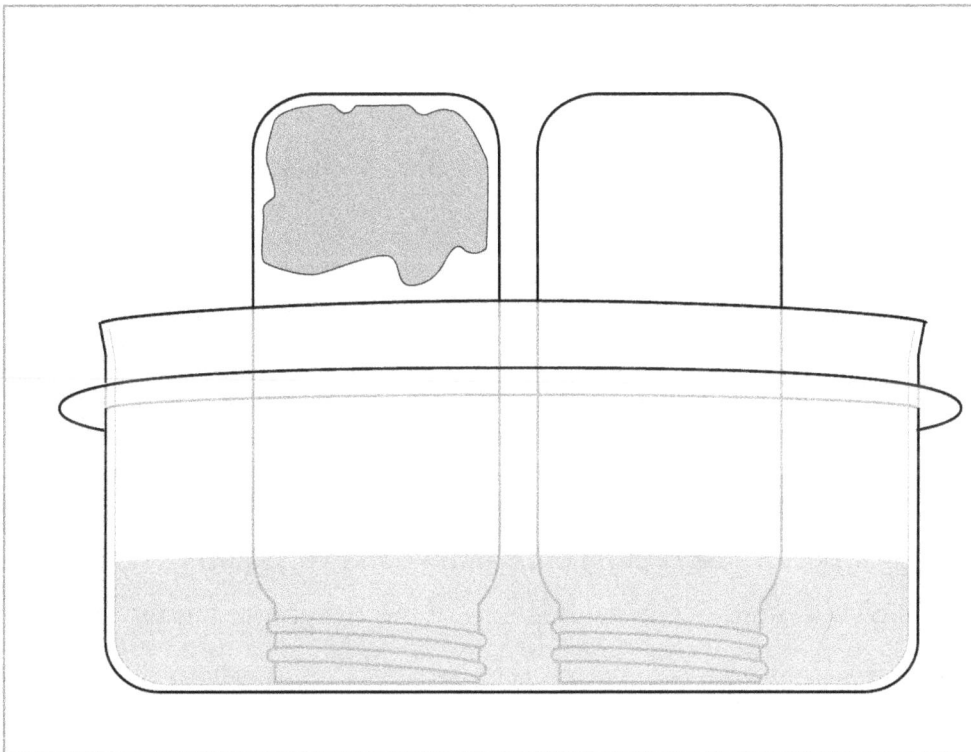

Figure 1.2 Schematic diagram

Data Table

Trial	Vol of 1st Jar	Vol of 2nd Jar	% O2
1			
2			
3			
4			
5			
6			

QUESTIONS:

1. Why is it important to have more than one trial for this experiment?

2. The actual percentage of oxygen in the atmosphere is ~20.9%. If the percentage you calculated is different, what things might account for the difference?

3. Sometimes a similar experiment is done in which a lit candle is used in the place of steel wool. The candle experiment can be done very quickly. However, the steel wool will give a more accurate measure of the oxygen in the air. Explain why this is the case.

Lab 3: Particles in the Air

INTRODUCTION

The atmosphere contains great numbers of suspended particles from natural sources, such as the oceans (salt), forests (pollen) and volcanoes (ash), from human inventions, such as motor vehicles, power stations and industrial plants, and from human activities such as agriculture. A combination of air and the suspended particles is called an *aerosol.* Some particles, liquid and solid, play important roles in cloud, rain, and snow formation. Other particles are hazardous to health and costly to society.

Since suspended particles are difficult to see it is surprising how many there actually are in the air. In polluted air over land there may be more than 1000 particles per cubic cm, while in clean air over the ocean the count may be as few as 100 per cubic cm.

ACTIVITY

OBJECTIVE: The objective of this activity is to investigate the amount and types of particulate matter in the air.

MATERIALS:
§ coffee filter
§ strainer
§ magnifying glass
§ 9-inch pie plate
§ white paper plate
§ filtered or distilled water

PROCEDURE:
Note: Some of this activity will be done at home.

1. Fill the pie plate full of water, and set it outside in a place that is exposed to the open air, and where the plate will not be disturbed. Leave the plate in place for 48 hours. If there is precipitation forecast in your area during the period of your experiment you can eliminate the distilled water and just filter the collected rainwater or melted snow water.
2. At the end of this time, take the coffee filter and put it in the strainer. If you have an automatic drip coffee maker, the filter holder will work best for this.
3. Take the filter and the strainer outside to the plate and carefully pour the water through the filter.
4. Rinse the sides of the plate with a small amount of water and pour this through the filter.
5. Carefully remove the filter and spread it out on the plate.
6. Allow the plate to sit undisturbed until the filter is completely dry. Be sure that there is no breeze near the filter while it dries. If there is a breeze, it will blow the particles away.
7. After the filter has dried completely, carefully fold it up and put it in an envelope. This will prevent the loss of particles while transporting it to class.
8. In class, use the magnifying glass to examine the particles on the filter. Without the magnifying glass, you may miss a lot of the particles.

QUESTIONS:
1. Describe what you see on the filter. What was your reaction when you made observations of the coffee filter? Were there more particles than you expected? Fewer? What about the types of particles; their colors, and sizes?

2. Compare your filter with your classmate's filter. *Explain some reasons why the particles on the filters may appear different.*

3. If there was precipitation in your area how might the results differ from those during a dry period?

4. Where do you think the different particles came from?

5. Some of the beneficial aspects of particles in the air have already been described. What harmful effects of these particles can you imagine?

6. What time of the year do you think you would have found more particles on the filter? Fewer? Explain

7. Are there areas of your city or town where you think particulate air pollution would be worse than where you did the activity? If so, where, and why do you think this?

8. Extra Credit: What weather conditions would increase the amount of particles in the atmosphere? Decrease the amount?

Lab 4: Introduction to Analyzing a Weather Map

INTRODUCTION

There are many types of weather charts or maps that the National Weather Service distributes to aid operational forecasters. One of the most important types is an *analysis* of the current weather showing observations collected by the global network of instruments. An analysis can depict many things related to the weather, including frontal positions and contours of constant pressure and temperature, etc. Another type of weather chart shows forecasts of the future state of the weather (called *prognoses* or *progs* for short) through a period of several days, and is generated by numerical weather predictions models run on super computers at the National Centers for Environmental Prediction located near Washington D.C. In this lab, we will concentrate on a surface analysis chart. Additional surface analysis exercises and explanations are provided in Lab 30.

Universal Time

Observations from around the country are plotted on weather maps, allowing the meteorologist to see weather patterns across the United States. Weather observations are always taken with respect to time. Thus all observations have a corresponding time, and that time is reported along with the measurement. By convention, atmospheric scientists use a twenty-four hour clock, and use one time zone, Universal Time (UTC), which refers to the time kept on the Greenwich meridian (longitude zero). This time is also known as Greenwich Mean Time (GMT) or Zulu time (Z). To convert to local time, we must know the time difference between UTC and local time for both standard time and daylight savings time. Not all locations use daylight savings time in the summer (e.g., Hawaii). Use of a universal system for measuring time allows observations to be made on a *synoptic* basis (from the Greek *synoptikos* meaning *general view of the whole*). Therefore, meteorologists refer to weather maps as *synoptic charts.* Synoptic charts are labeled with the valid time and the atmospheric level (surface vs. upper air) of the data plotted on the chart.

Station Model Explanation

Data on a surface analysis are plotted in a uniform manner around the point locating the observing station on the map (Fig. 1.3). Once the observations are plotted on the map, meteorologists analyze the map to become familiar with current patterns of low and high pressure, temperature, dew point, etc. Contours are drawn for lines of constant pressure, called *isobars,* and lines of constant temperature, called *isotherms*.

The standard convention for plotting pressure is to plot it in tenths of a millibar (mb) and drop the leading number. For example 1011.1 is plotted as 111 and 996.7 is plotted as 967. For the exercises in this book pressure is rounded to the nearest mb and plotted in its entirety; 1011.1 is plotted as 1011, and 996.8 as 997. This simplification is undertaken to make the task of isobar analysis more transparent. However, the standard convention should be expected in most cases when encountering plotted surface pressure data.

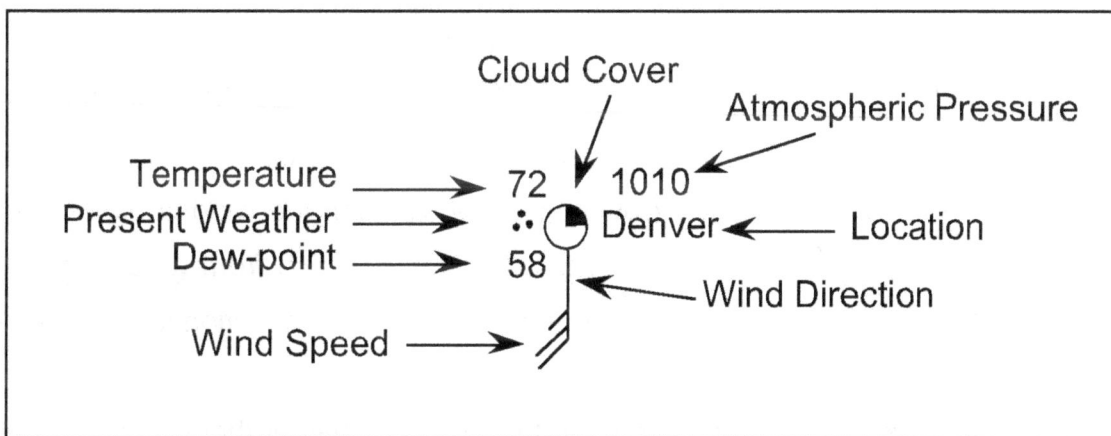

Figure 1.3 Station Model for plotting of surface weather observations

Cloud Cover - This represents the amount of cloud cover over the station. One quarter of the circle is filled in for one-quarter cloud cover, one half for one half-cloud cover, etc. If the sky is obscured an X is put in the circle.

Atmospheric Pressure - This is the atmospheric pressure measured in millibars (mb). Meteorologists draw contours of equal pressure called *isobars*. In practice the number recorded on the station model includes only the last two numbers and the decimal place. However, for this lab, the entire pressure is plotted in whole millibars.

Temperature - This is the temperature measured in °F at the top of each hour. Isotherms designate areas of equal temperature.

Dew-point Temperature - This is the dew point temperature measured in °F at the top of the hour. You will recall that the dew point temperature is the temperature at which water droplets form.

Wind Direction - This line represents the direction from which the wind is blowing.

Wind Speed - The small barbs represent the wind speed. Each full line represents ten knots (1 kt =1.15 mph). Shorter lines represent wind speed increments of 5 knots. If the winds speed exceeds 50 knots, a triangle shaped barb is used. The total wind speed is determined by adding the barbs (See Fig. 1.4).

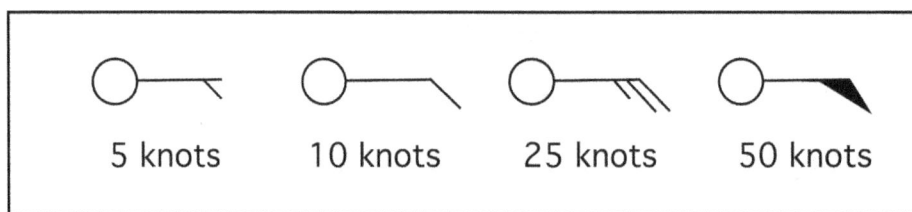

| 5 knots | 10 knots | 25 knots | 50 knots |

Figure 1.4 Convention for plotting wind speed and direction.

If you think of the wind speed barbs as feathers on an arrow, the circle represents the arrowhead. The arrow points the direction the wind is blowing to. In meteorology the wind direction is designated as the direction from which the wind is blowing. Therefore, if an arrow points to the *west*, the wind direction is actually called *east* (Fig. 1.4).

Present Weather - Symbols are used to show the weather that is occurring at the time of observation (Fig. 1.5).

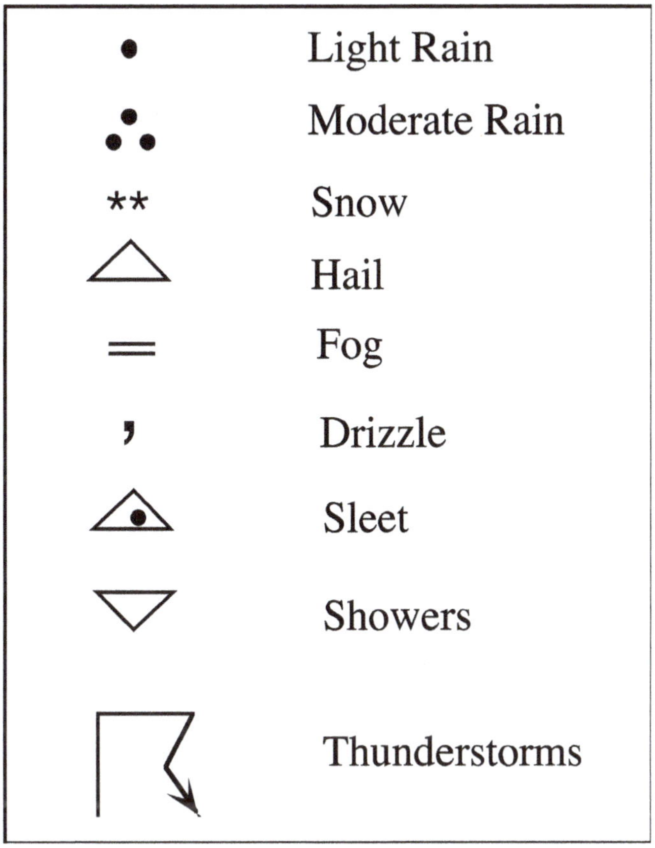

●	Light Rain
∴	Moderate Rain
**	Snow
△	Hail
=	Fog
'	Drizzle
△̇	Sleet
▽	Showers
⌐↘	Thunderstorms

Figure 1.5 Current weather symbols

Charts that depict the current conditions in the atmosphere above the surface are also very valuable to the forecaster, and are similar to surface maps in that they show observations plotted for a particular time. Surface maps are issued every three hours, whereas upper-air maps are issued only twice daily corresponding to the times at which radiosonde balloons are launched. In addition to data collected by radiosonde instruments, upper air maps may also show aircraft reports and satellite data.

ACTIVITY

OBJECTIVE: This activity provides an introduction to the art of analysis and interpretation of weather data.

PROCEDURE:

1. Analyze the data provided below by drawing contours with values of 3 and 8.

10	10	10	10	10	10	10
5	5	5	5	5	5	5
1	1	1	1	1	1	1

2. Analyze the data provided below by drawing contours with values of 3, 6, 9, 12, 15, and 18. Label your contours.

10	10	10	13	15	20	21
5	5	10	12	13	15	18
1	2	4	8	10	13	14
1	1	4	7	8	9	11

3. Analyze the data provided below by drawing contours with values of 120, 140, 160, and 180. Label your contours.

100	100	100	100	100	100	100
100	110	120	150	120	110	100
100	120	150	180	150	120	100
100	150	180	200	180	150	100
100	140	170	190	170	140	100
100	120	150	180	150	120	100
100	100	100	100	100	100	100

4. Decode the following observation.

5. Analyze the temperature data provided in Fig. 1.6 by drawing contours of constant temperature (isotherms) with values of 40, 50, 60, and 70° F. Label your contours.

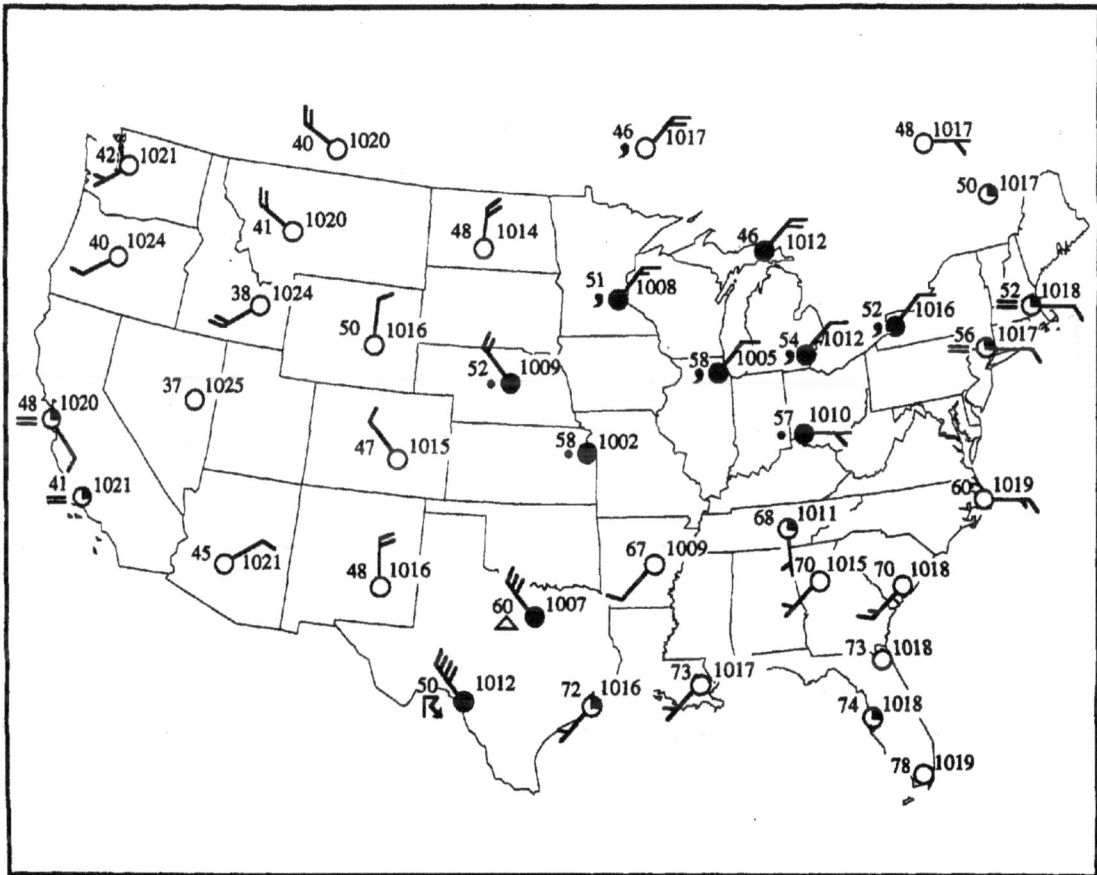

Figure 1.6 Plotted weather map

6. Analyze the pressure data provided below by drawing contours of constant pressure (isobars) with values of 1004, 1008, 1012, 1016, 1020, and 1024. Label your contours.

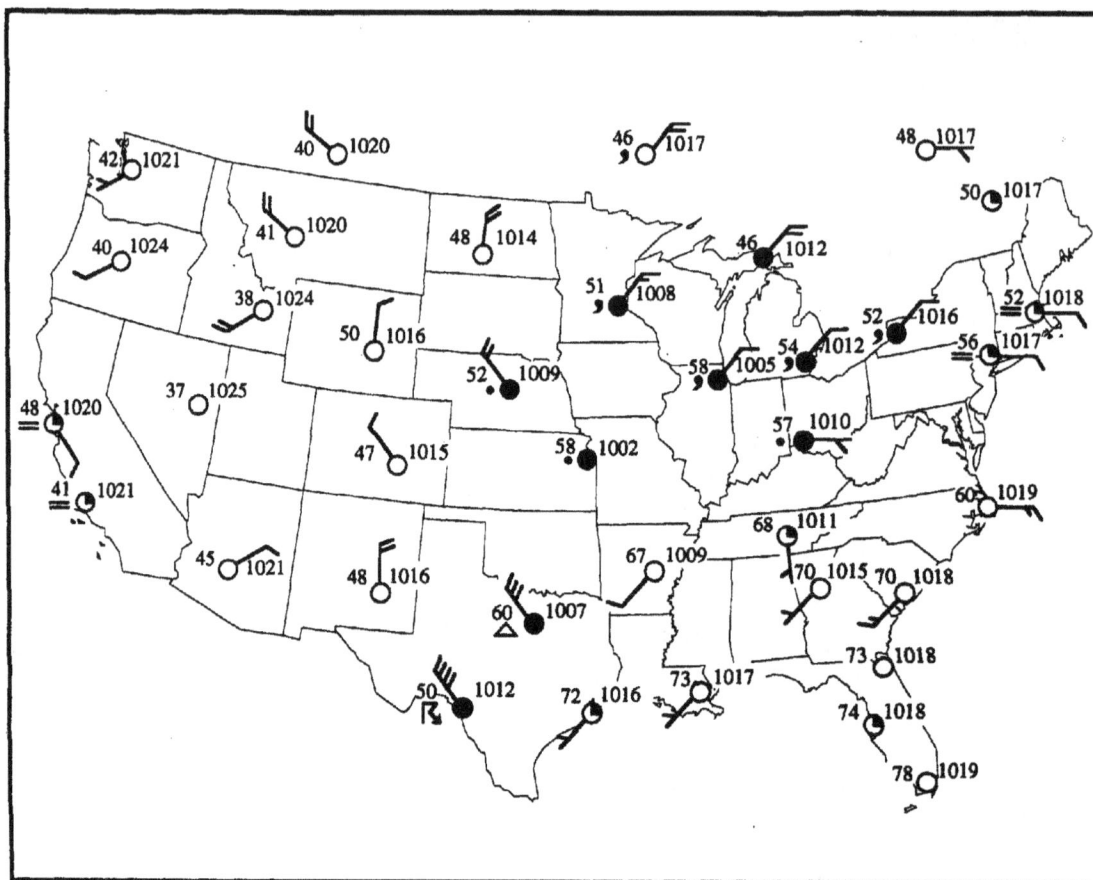

Figure 1.8 Plotted weather map

7. Qualitatively, what is the relationship between your isobars above and the wind direction shown by the station plots?

Lab 5: Indirect Measurements

INTRODUCTION

There are a number of instances in meteorology where it is important to know the distance to objects that are beyond the reach of a ruler. For example, in the next section the distances from the sun to various planets in our solar system are given. How do we make measurements of these distances? Unlike objects on Earth, we cannot hold ruler up to the planets and measure them. There are, however, indirect methods to make measurements of remote objects. In this lab we use angular diameter to make indirect measurements of objects. This technique can be applied to measuring the diameter of the moon or the height of distant clouds or trees.

Meteorologists apply angular diameter to gauge the height of clouds from radar signals returned as echoes by cloud particles. The time it takes for the echo to return provides the distance to the cloud and the clouds angular diameter as seen by the radar can then be used to calculate the height of cloud top. Weather forecasters use this information in determining the hazards represented by storms. The higher the cloud top of a thunderstorm, the more likely it will produce severe weather such as large hail or tornadoes (see Chapter 8).

The easiest way to understand an angular diameter is to look at an example. For instance, it is possible to hold a penny close enough to your eye so that it just blocks out the head of one of your classmates. You have adjusted the position of the penny so that it has the same angular diameter as your classmate's head even though the penny is much smaller. This example shows why angular diameters are useful in making measurements. If the true diameter of the penny is known, the true diameter of the person's head can be mathematically determined without measuring it directly. Many objects in our environment can be measured the same way. The following activity will demonstrate how this principle can be used to determine the height of objects on your campus, the diameter of the moon (The angular diameter of the moon is 0.5°), and the height of clouds.

ACTIVITY

OBJECTIVE: The objective of this activity is to learn how angular diameters can be used to make an indirect measurement of the true height of objects.

MATERIALS:

For each student or group of students:

§ paper plate

§ meter stick

§ construction paper

§ pencil (round ones work best)

§ two pushpins

PROCEDURE:

1. Make a pinhole in the construction paper and attach it to the end of the meter stick with push pins or tacks.
2. Attach the paper plate to the wall in a place so that you can stand exactly 8 meters from it as shown in Fig. 1.9. Attach the plate at eye level or a little higher.
3. Point the meter stick at the paper plate and look through the pinhole along the meter stick at the plate.
4. Slide the pencil along the meter stick until the thickness of the pencil just covers the paper plate. The pencil and the plate now have the same angular diameter (Fig. 1.9).
5. Read the distance from the pinhole to the pencil and record it in the Data Table. Add 1 cm to this measurement to account for the fact that the meter stick was not exactly against your eye.
6. Repeat steps 3 through 5 three more times recording each measurement in the Data Table. Then calculate the average of the four measurements and record it in the Data Table as well.

7. Using the meter stick, measure the diameter of the pencil and the distance to the plate and record them in the Data Table. Make sure each measurement is in millimeters.

8. Using the equation below, calculate the diameter of the plate and record it. This is possible because even though the pencil and plate are different sizes, the pencil was moved to the point where it had the same angular diameter as the plate.

Diameter of the plate = (distance to the plate) X (diameter of the pencil)

(distance to the pencil)

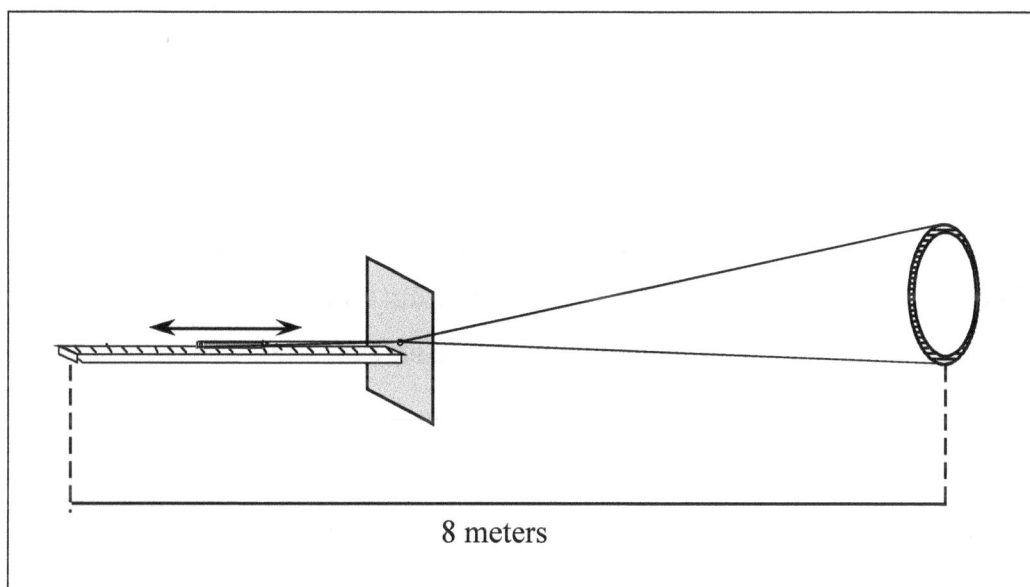

Figure 1.9 Schematic diagram

9. Measure the true diameter of the plate and record it in the Data Table.

10. Try this activity to determine the height of objects on your school campus. Everything is done the same way except you will measure the height of the object instead of the diameter of a plate. You will need to know the distance to the base of your chosen object; a tree or building, and then follow the preceding steps.

11. On a clear night with a full moon this procedure can be used to calculate indirectly the diameter of the moon. To do this you follow the steps above and use the fact that the distance to the moon is ~ 384,401,000,000 mm.

QUESTIONS:

1. Compare the true diameter of the plate with the diameter you calculated. How do the two compare?

2. If the true and calculated diameters are not the same, what reasons could explain the difference?

3. Would this method of determining diameters be helpful in measuring the diameters of the planets? Why?

4. Extra Credit: The full moon appears to be bigger when it is on the horizon than when it is high up in the sky. How could you use the method in this activity to determine whether this appearance is real or an illusion?

Data Table

	1	2	3	4	Average
Distance from pinhole to pencil (cm)					
Diameter of pencil (cm)					
Distance to plate (cm)					
Diameter of plate (cm)					
True diameter of plate (cm)					

$$\boxed{} = \frac{\boxed{} \times \boxed{}}{\boxed{}}$$

Chapter 2 Pressure And Temperature

'All truth passes through three stages. First, it is ridiculed. Second, it is violently opposed. Third, it is accepted as being self-evident.' – Arthur Schopenhauer

Pressure and temperature are two of the fundamental variables used to describe the state of the atmosphere. Their vertical distribution in the atmosphere was mentioned briefly in Chapter 1. The relationship between pressure and temperature is important in understanding atmospheric motions, storm systems, and climate. Therefore a short summary is presented here.

Pressure

As stated in Chapter 1, atmospheric pressure is the weight exerted by the overlying air molecules in a column of unit cross section extending upward to the top of the atmosphere. Pressure can also be described as the force per unit area exerted by the continuous collisions of gas molecules. How are these two dissimilar descriptions reconciled? The first one looks at pressure from a larger perspective: weight is a force resulting from the acceleration due to gravity acting on the mass of the overlying air. The latter description is from a molecular perspective: air pressure is proportional to the *speed and mass of the air molecules*, and the *frequency of their impacts*. Since the force due to atmospheric pressure is the same on every part of your body, you don't feel it as weight. This fact led the ancients to hypothesize that air had no weight at all.

The weight of the atmosphere can be measured by means of a *barometer* (first invented by Torricelli in 1643). Liquid mercury is carefully poured into a "J"-shaped glass tube and then the tube is inverted, creating a vacuum space at the top of the tube. The height to which the mercury rises in the tube due to the weight of the atmosphere is a measure of the atmospheric pressure. If one were to carry a barometer up the side of a mountain, the height of the mercury column would gradually decrease, reflecting the fact that there is less air above the level of the barometer as its elevation increases.

Temperature

Temperature can be thought of as the degree of hotness or coldness of an object. *Temperature is proportional to the speed and mass of the air molecules* (their average kinetic energy). From the molecular view it is clear that there must be a close relationship between pressure and temperature, since pressure is also related to the speed and mass of air molecules, as well as the frequency of their impacts. A simple experiment can be conducted to clarify this relationship. Take an air-tight container (constant volume of air) and insert a thermometer and a pressure gauge. Then place the container on a source of heat (e.g., burner). We find that as the temperature rises the pressure also rises. When a graph is made of the pressure versus the temperature the result is a straight line (a linear relationship between pressure and temperature).

When this straight line on the graph is extended to the pressure-axis where pressure equals zero, the coldest temperature theoretically possible is reached. This temperature is referred to as zero *Kelvin* (in honor of the scientist who made this discovery). Kelvin units are abbreviated K. Scientists use a temperature scale starting with zero K, which eliminates negative temperatures, for all calculations relating temperature and pressure. Thus, in calculating pressure from temperature using Kelvin, the result will always be greater than zero, consistent with the observation that a negative pressure value is without meaning.

Two other commonly used scales are the Fahrenheit scale and the Celsius scale. The Fahrenheit scale was named for G. Daniel Fahrenheit, the scientist who developed it. The value of 32 was assigned to the temperature at which water freezes and the value of 212 was assigned to the temperature at which water boils. The value of 0 was the lowest value Fahrenheit obtained in his experiments with a combination of salt water and straw. Fahrenheit used the human body temperature of near 100°F (98.6°F) to fix a second point on his scale. Between the freezing point and the boiling point are 180 equal divisions, each called a degree.

The Celsius scale was developed after the Fahrenheit scale, but in a similar fashion. A Swedish astronomer, Anders Celsius, assigned the values of 0 and 100 to the freezing and boiling points of water. The distance between the two points was divided into 100 equal increments, also called degrees. Each degree Celsius is 1.8 times larger than a degree Fahrenheit.

The thermometer as a meteorological instrument has a long history. A Greek scientist named Hero first constructed an early version about 2000 years ago. A water thermometer was invented in 1597 by Galileo, which must have broken with every freeze. Later wine and mercury were used as the expanding liquid in the thermometer.

Temperature variations across the Earth's surface are also associated with differential heating (land, sea), altitude, latitude (geographic position), and ocean currents. Contours of constant temperature on weather maps are referred to as isotherms. The subject of winds will be discussed further in Chapter 6.

Heat

Heat is that which makes things hotter, the transfer of energy from one object to another. Heat is measured in units called *calories*. A calorie is the amount of heat (or energy) that is required to raise one cubic centimeter of water by one degree Celsius. For a given substance, the temperature change caused by any quantity of heat depends on the mass of material involved and the *specific heat*. The specific heat is the quantity of heat that must be added to a unit mass of a substance in order to raise its temperature 1°C. Thus, the specific heat of water is 1 cal/gm °C (or the magnitude of the specific heat of water is 1).

There are four ways in which heat is commonly transferred in the atmosphere. *Conduction* is the transfer of heat between molecules within a substance. When you leave a metal spoon in a pot of boiling water, the spoon becomes hotter. This is an example of conduction. Wind chill is an example of conduction in the atmosphere.

Another form of heat transfer is *convection*. The transfer of heat by the mass movement of a fluid is called convection. Convection occurs mostly in liquids and gases. When a pot of water is boiling, convection is occurring. In the atmosphere, thunderstorms are a vigorous form of convection that transfers heat and moisture from the surface into the atmosphere.

A third form of heat transfer is *radiation*. A large amount of heat is received from the sun in the form of radiation. Energy travels from the sun to the Earth in the form of electromagnetic waves, which are largely absorbed at the Earth's surface. Heat from a fire is another example of energy transfer by radiation. An expanded discussion of radiation is given in Chapter 3.

The last method in which heat is transferred in the atmosphere is often overlooked. It is the *latent heat* associated with changes in phase of water. For example, heat energy is required to evaporate water. The energy in the form of latent heat is released in convective updrafts when water vapor condenses into cloud droplets. This added heat increases the buoyancy of clouds, which acts as a driving force for thunderstorms. An expanded discussion of latent heat is given in Chapter 4.

Lab 6: Atmospheric Pressure

INTRODUCTION

Contrary to our intuition, air pressure is exerted equally in all directions: down, up, and sideways. Therefore, when you are inside a building, air pressure is not just the weight of all of the air within the room up to the ceiling, but it is the weight of the air in a column above the building that extends to the top of the atmosphere.

ACTIVITY

OBJECTIVE: The purpose of this activity is to demonstrate the nature of atmospheric pressure.

MATERIALS:
§ stiff plastic cup or a glass
§ stiff paper cup
§ cardboard or other stiff card
§ straight pin
§ water

PROCEDURE:
1. In doing this activity, work over a sink or catch basin. Fill the plastic cup (or glass) up to the rim with water. Cover the cup with the cardboard. *In the space below, predict what will happen if you turn the cup over with the cardboard covering the cup, and explain your prediction.*
2. While holding the cardboard onto the cup, carefully turn the cup over. Hold the cup by the bottom and release the cardboard as shown in Fig. 2.1. In doing this part of the activity it is important that the cup is not deformed in the process of turning it over. Why is this so?
3. Slowly, turn the cup sideways. *What happens?*

4. Use the straight pin and carefully make a hole in the bottom of the cup. Then carefully repeat the process while holding a finger over the hole in the bottom of the cup. Lift your finger. *Explain the result.*

Repeat the same process with the paper cup. *Were the results any different?*

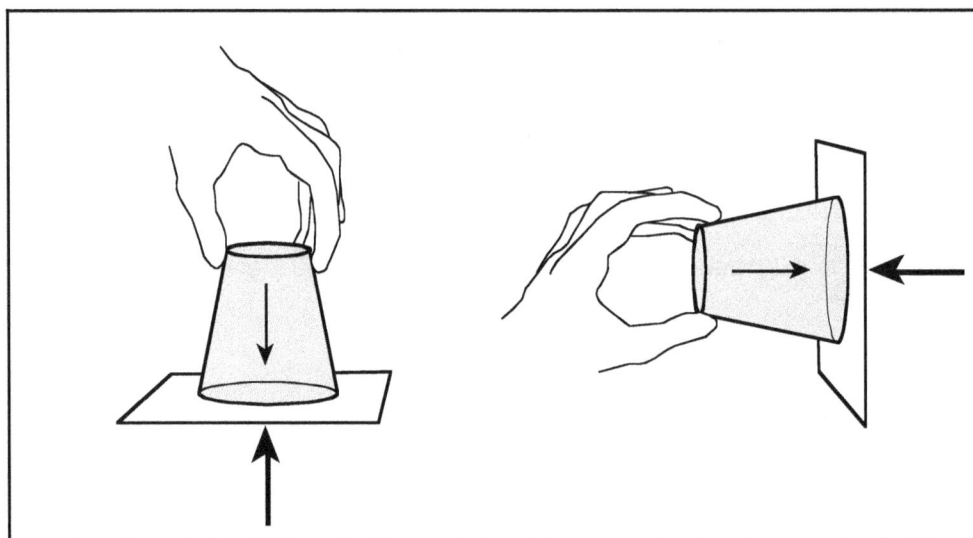

Figure 2.1 Shifting from the position on the left to the position on the right must be done slowly and over a sink rather than a student's head.

QUESTIONS:

1. In this activity, what held the cardboard to the cup, preventing the water from falling out of the cup?

2. Explain why the water and the cardboard fell from the cup when your finger was lifted from the pin hole in the bottom of the paper cup.

Lab 7: Crushing Cans

INTRODUCTION

This activity provides a demonstration of the force that is exerted by atmospheric pressure. This experiment works well as a demonstration. If done correctly, it will produce dramatic effects. If, however, the experiment fails, the activity is easily and quickly repeated. (This activity can also be performed using a larger gasoline can – preferably without the gasoline.)

ACTIVITY

OBJECTIVE:

The purpose of this lab is to explore the relationship between pressure and temperature, and the magnitude of the force represented by atmospheric pressure.

MATERIALS:
 § aluminum can
 § pan
 § tongs
 § water
 § gas burner

PROCEDURE:
1. Put cold water and ice in the bottom of the pan as shown in Fig. 2.2.
2. Put a tablespoon or so of water in the bottom of the aluminum can and heat the can until a cloud appears. The can needs to be rather hot for this experiment to work well (use the presence of steam as a guide). **CAUTION: *do not*** touch the can with your bare hands.
3. When wisps of cloud* appear at the opening of the can, use the tongs and quickly invert the can in the pan of water. (*Sometimes people mistakenly refer to this cloud as steam. Actually steam is invisible. It is not until the vapor cools and

condensation takes place that a cloud becomes visible. See section 6 for more on this subject)

4. Repeat steps 1 through 3, using water at room temperature in the pan. Can you predict the outcome of the modified experiment?

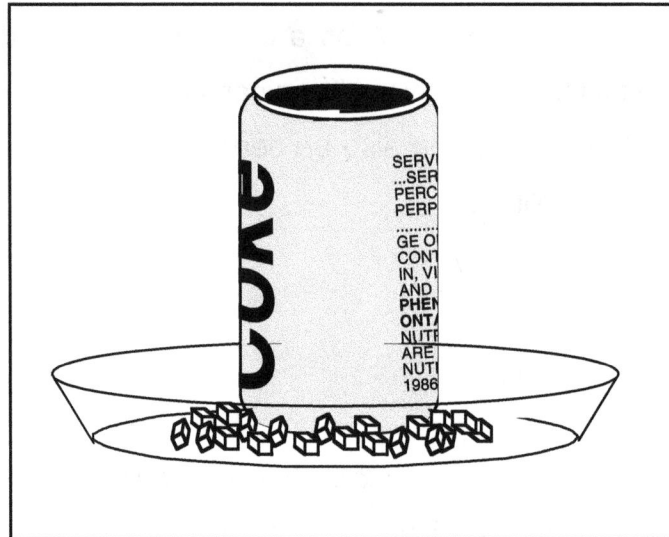

Figure 2.2 Schematic diagram

QUESTIONS:

1. What happened to the can in this activity? Explain.

2. When the can emerges from the water, it is practically filled with water. Explain why.

3. How did the result differ when water at room temperature was used in the pan? Explain.

4. The tablespoon of water placed in the can in step 1 plays an important role in this experiment. Explain.

Part II: Toss a piece of burning notebook paper into a Pyrex flask or jar with a mouth slightly smaller than an egg. Place the peeled egg on top of the flask's mouth and carefully watch the egg when it is first placed on the flask (Fig. 2.3).

Figure 2.3 Schematic diagram

1. When does the egg enter the flask?

2. Why does the egg enter the flask?

3. What will happen if the flask is inverted and heated gently with a candle? Why?

Lab 8: Exploding Popcorn

INTRODUCTION

Dramatic pressure changes occur when water changes phase from liquid to vapor. This is one of the reasons why water boils at 100 °C at sea level. At this temperature the vapor pressure of water is the same as the atmospheric pressure at sea level. As more heat is added to the pot of boiling water, the temperature of the water stays at 100 °C. The added energy goes into breaking the bonds between the water molecules as they are released as individual vapor molecules. However, if the pot has a tight lid (as in the case of a pressure cooker), the temperature of the water can continue to rise, and consequently the food will cook faster. Eventually, the pot would explode if enough energy was added and it did not have a safety valve to release the pressure. This is what happens in the case of some volcanic eruptions, such as Mount St. Helens (see Lab 46). Water, from snowmelt, percolates into the neck of the volcano where it is heated by the hot magma within. The overlying rock acts as a lid, which finally is displaced by a small earthquake or tremor, releasing a tremendous built-up of pressure in a catastrophic explosion.

A single kernel of popcorn can provide some interesting pressure physics to demonstrate these principles. A popcorn kernel is made up of starch, protein, fat, minerals, and water. The presence of liquid water is critical to popping. When popcorn is heated, the water inside the kernel becomes vapor and the internal pressure increases to as much as 9 atmospheres (1 atmosphere ~ mean sea level pressure). When the hull can no longer withstand the pressure difference between the pressure inside and the air pressure outside, the kernel explodes or pops.

ACTIVITY

OBJECTIVE: To determine the moisture content of popcorn kernels and to observe how moisture and pressure relate to the popcorn's ability to pop.

42

EQUIPMENT/MATERIALS:

§ normal popcorn

§ oven-heated popcorn

§ 125-ml Erlenmeyer flask

§ aluminum foil

§ Bunsen burner

§ ring stand and ring

§ wire gauze

§ ruler

§ permanent felt marking pen

§ balance

§ 10-ml graduated cylinder

§ heat-resistant glove or flask tongs

§ 100-ml graduated cylinder

PROCEDURE:

1. Using aluminum foil, fashion a lid for a 125-ml flask. Determine the mass of the empty flask and lid. Record it in the following Data Table. Be sure to put units on the data.

2. Using the 10-ml graduated cylinder, measure the volume occupied by 20 kernels of unpopped normal popcorn. Record your findings in your Data Table.

3. Put the 20 kernels of unpopped corn in the Erlenmeyer flask and determine the mass of the flask, popcorn and the foil lid. Record this information.

4. Using subtraction, determine the mass of the 20 kernels of unpopped corn. Record this data.

5. Make sure the foil lid fits snugly over the mouth of the flask. With a pencil or a pen, make a few small holes in the foil.

6. Heat the flask over the Bunsen burner to pop the corn. Be sure to use wire gauze between the flask and the burner. Let the flask sit undisturbed until the kernels start to change color. Then use flask tongs or heat-resistant gloves to hold the flask. Shake the flask over the burner until all kernels pop.

7. Remove the flask from the heat and carefully remove the foil. CAUTION! Do not get burned by the steam! Let the flask stand for a few minutes to cool.

8. When the system is cool, determine the mass of the flask, lid and popped corn. Record this information.

9. Mark a kernel of oven-heated popcorn with a felt marker and place it in a 125-ml Erlenmeyer flask along with a normal kernel. Cover the flask tightly with a foil lid. Heat the flask as instructed before. Watch the two kernels to determine which pops first and which produces the fluffiest popcorn. Record these observations.

Measure in millimeters the longest side of each of the two popped kernels. Record this information.

DATA TABLE

MASS OF EMPTY FLASK AND LID	
VOLUME OF 20 UNPOPPED NORMAL POPCORN	
MASS OF FLASK, LID, UNPOPPED POPCORN	
MASS OF UNPOPPED POPCORN	
MASS OF FLASK, LID, POPPED CORN	
FIRST TO POP: HEATED OR NORMAL?	
FLUFFIEST: HEATED OR NORMAL?	
LONGEST SIDE OF HEATED (mm)	
LONGEST SIDE OF NORMAL (mm)	

QUESTIONS:

1. Why is the mass in popped corn less than the mass in the unpopped corn (i.e., what has left the system)?

2. What is the mass of the water lost by the system? Show mathematically how you arrived at this answer.

3. What was the percentage of water in the unpopped kernels? Show mathematically how you arrived at this answer.

4. The ratio of the volume of the popped corn to the volume of the unpopped corn is called expansion volume. The popcorn industry uses expansion volume as a test of quality. Orville Redenbacher claims his gourmet popping corn has an expansion volume of 40 to 1. What is the expansion volume (expansion ratio) for the popcorn used in this experiment? Show mathematically how you arrived at this answer.

5. What substance is present in a larger amount in the normal kernel as compared with the oven-heated kernel? In which kernel was the internal pressure greater before popping?

6. Which kernel produces more desirable popcorn, the normal or the oven-heated?

7. Reread the lab introduction and answer the following questions: Would popcorn pop faster on Pike's Peak or in Boston? Why?

8. Prepare a graph of expansion volume (dependent variable with no units) as a function of the percentage of water (independent variable with % units). A Data Table must accompany the graph. Use: Expansion Volume, Percentage Water (%) and Source of Data, as column titles. The Source of Data column should identify the lab partnership from which the data was obtained.

9. Use data from the whole class and make the graph cover as much of a piece of graph paper as possible. Draw a best-fit line on your graph. *Remember:* Label the axes of the graph and show the units where appropriate. Protect the points on your graph with circles, squares or triangles. Title and number your graph.

Lab 9: Density, Temperature, and Pressure

INTRODUCTION

In Lab 7: Crushing Cans, a closed container (constant volume) was used. In the open atmosphere the relationship between air pressure, temperature, and volume is somewhat more complex. When the temperature of the air changes, so does its density. Molecules move closer together or farther apart as the temperature increases or decreases. Therefore, when air is heated it expands, becoming less dense than the surrounding air and rising relative to the surrounding air. Conversely, when air is cooled it contracts, becoming denser than surrounding air, and therefore it sinks.

Within the atmosphere, when air temperature is increased, the expanding air mass rises, creating a region of low surface pressure. When air temperature is decreased, the contracting air mass sinks, creating a region of high surface air pressure (recall that surface pressure is just the total weight of all the overlying air molecules. Since colder air molecules take up less space than warm ones, the total number of overlying air molecules in a cold air mass is larger and the surface pressure is higher). This relationship between air temperature and the air pressure in an air mass is counter to the result of the closed container experiment; a potential a source of confusion.

ACTIVITY

OBJECTIVE: The objective of this activity is to investigate the relationship of temperature and density.

MATERIALS:
§ balloon
§ empty 750-ml bottle
§ bucket of ice
§ bucket of hot water (start with water heated to near boiling)

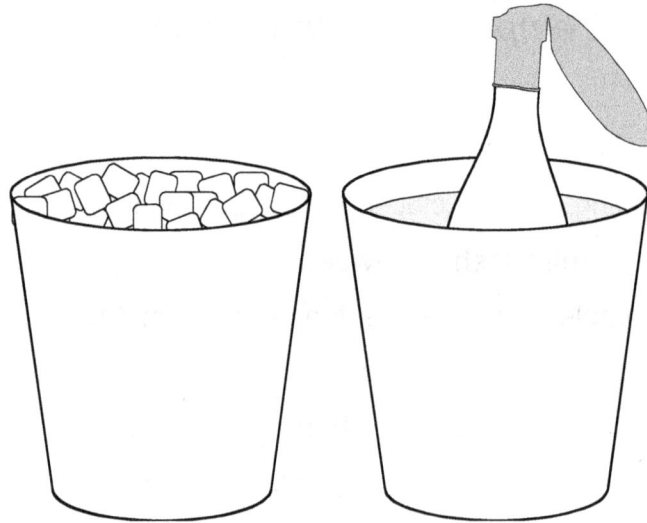

Figure 2.4 Schematic diagram

PROCEDURE:

1. Place the uncovered bottle in the bucket of hot water for three minutes (Fig. 2.4). Caution should be used when handling very hot water. Do not submerge the bottle or allow water to get into the bottle.

2. Place the balloon over the mouth of the bottle. You now have an isolated mass of air. It is important to remember throughout this experiment that the amount, or mass, of air will remain constant. In the space below, predict what will happen to the balloon if the bottle is placed in a bucket of ice water. Explain your prediction.

3. Place the bottle in a bucket of ice water for three minutes. What happens? Remembering that the mass of the air has remained constant, what has changed?

4. Take the balloon off the bottle and place the bottle back in the bucket of ice for three minutes. Do not submerge the bottle or allow water to get into the bottle.

5. Place the balloon over the mouth of the bottle. As before, you have an isolated air mass. In the space below, predict what will happen to the balloon if the balloon is placed in a bucket of hot water. **Explain your prediction.**

6. Place the bottle in a bucket of hot water for three minutes. What happens? Remembering that the mass of the air has remained constant, what has changed?

QUESTIONS:

1. What are the two variables of air that are being changed in this activity? What is the relationship between these two variables?

2. As the air is heated, what affect does this have on the air molecules?

3. In the atmosphere, what would you expect to happen to air that is warmed? Cooled?

4. Based on your observations and your answers to these questions, do you think it would be best to place a warm air vent near the floor or ceiling of a room? Explain your answer.

5. Extra Credit: As a balloon rises it expands and the air inside it cools. Explain why these observations are not contradictory.

Chapter 3 Sunlight and the Earth's Climate

'In science one tries to tell people, in such a way as to be understood by everyone, something that no one ever knew before. But in poetry, it's the exact opposite.'
- Paul Dirac (1902-1984)

In 1633, the famous Italian scientist, Galileo Galilei was jailed by Church leaders for his efforts to persuade his countrymen that the Earth moves around the sun. The Church has since agreed that it does indeed do so in a nearly circular orbit with a period of about 365 days. At the same time, the Earth revolves about its axis once every 24 hours. The Earth's equatorial plane is tilted at an angle of 23.5° with respect to the plane of the orbit. Because of this tilt, the Northern Hemisphere intercepts maximum quantities of solar radiation during the middle months of the year, May, June and July, and the Southern Hemisphere experiences maximum solar radiation during the months of November, December, and January. The length of the day, the length of the path that sunlight must traverse through the atmosphere, and the surface area over which the radiation is distributed all vary with latitude and season and influence the amount of sunlight available to warm the Earth. Galileo reasoned that the seasons of the year are clearly related to these observations.

Radiant energy from the sun is in the form of *electromagnetic waves* and travels at the speed of light, 3×10^8 meters per second (m/s). The wavelengths of the sun's radiation cover a wide spectrum. Of particular importance in the atmospheric sciences are the ultraviolet band (wavelengths 0.01 to 0.4 micron or μm), the visible band (0.4 to 0.7 μm), and the infrared band (0.8 to 1000 μm). Most solar energy is found in the visible and infrared bands. One micron is one millionth of a meter – 1 μm = 1/1,000,000 meters.

The characteristics of the energy radiated by any body can be stated in three basic laws, which are derived from the more general Planck's law:

(1) *Stefan-Boltzmann law* - All objects emit radiant energy (except at 0 K), with hotter objects emitting more energy per unit area than colder objects.

(2) *Kirchoff's law* - Objects that are good absorbers of radiation are good emitters of radiation.

(3) *Wien's law* - The hotter the object the shorter the wavelength of the maximum emitted radiation.

The Stefan-Boltzmann law shows that the quantity of energy emitted varies with the fourth power of the absolute temperature, $E = cT^4$, where c is a constant. Therefore, if one body has a temperature of 6000 K (the sun) and another body has a temperature of 300 K (the Earth), the first one radiates $(6000/300)^4 = 160,000$ times more energy than the second for each unit area of surface.

A body that radiates energy according to the Stefan-Boltzmann law is called a blackbody. Most substances do not radiate as much energy as a blackbody does, since this represents a theoretical upper limit. The ratio of actual emission to blackbody emission is called the emissivity. It depends on the nature of the substance and may vary with wavelength and, to a lesser extent, with temperature. For example, snow has a high emissivity at infrared wavelengths, but a low emissivity at visible wavelengths.

A radiation law known as Kirchoff's law states that for any wavelength a substance's emissivity equals its absorptivity. The ratio of the actual energy absorbed to the total amount of energy intercepted by the substance is known as absorptivity. Kirchoff's law indicates that the absorptivity of snow is high at infrared wavelengths, but low at visible wavelengths. This accounts for the fact that snow reflects sunlight effectively and looks white.

The wavelength L at which the emission of radiant energy is a maximum (L_{max}) can be calculated from Wien's law. Specifically, the law is $L_{max} = 2,880/T$, where T is the absolute temperature and L_{max} is in micrometers (µm). Taking the sun's temperature at 6000 K, Wien's law indicates that maximum solar radiation is at a wavelength of 0.48 µm, which is in the visible band.

The quantity of radiant energy incident on the top of the Earth's atmosphere is called the *solar constant* and amounts to ~1400 watts per square meter (~2 calories per cm^2 per minute). The absorption, reflection, and scattering of solar radiation as it passes through the atmosphere depend on the wavelength. Most of the ultraviolet radiation is absorbed in the upper atmosphere. It leads to photo-ionization and to ozone formation and destruction.

Most of the visible radiation passes through the atmosphere and warms the Earth's surface. Some visible radiation is reflected from cloud tops and scattered by air molecules and atmospheric particles. The fraction of incident visible solar radiation that is reflected back to space is called the Earth's *albedo.* The albedo is difficult to measure accurately but is considered to average ~30 percent.

The *greenhouse effect* is the term used to describe the role of gases such as water vapor, carbon dioxide, and methane in increasing the temperature of the Earth's surface. These gases permit solar energy to pass rather freely through the atmosphere to the ground. But the infrared radiation emitted by the Earth is partly absorbed by these gases and prevented from escaping directly to outer space. Downward directed infrared radiation from greenhouse gasses in the atmosphere adds to the radiation budget at the ground, causing a warmer surface temperature than without an atmosphere. It is important to note that water vapor is the principal greenhouse gas.

An examination of the radiation budget of the Earth shows that at the surface the amount of radiation absorbed exceeds the amount emitted at tropical latitudes. The net radiation, or radiation balance as it is called, is positive. Conversely, the radiation balance at polar latitudes is negative; more radiation is lost to space than is available from sunlight. This radiation imbalance must be compensated for through other means of heat transport, or the poles would continue to cool and the equator would continue to warm.

The major mechanisms for heat transport from low to high latitudes are air and ocean currents. Much of the transfer occurs as warm air and water move poleward while cold fluids move southward. A substantial quantity of heat is transported as latent heat. As water is evaporated at low latitudes, heat is added to the atmosphere in latent form. It is released as heat in storm systems when condensation leads to clouds and precipitation. Latent heat is explored in detail in Chapter 4.

Air temperatures at the Earth's surface depend on various factors, including the radiation balance. In any particular region, the albedo of the surface is an important variable. A land area composed of highly absorbing material will have a lower albedo than a sandy or snow-covered area. The albedo of water depends on the altitude of the sun. The average albedo of the ocean is about 8%.

The specific heat of water is much greater than that of land material, such as rock and sand. Thus, for the same mass, land regions undergo greater temperature changes than water regions when equal amounts of heat are added or taken away.

Radiant energy falling on land is absorbed in the top few millimeters and is slowly conducted downwards. Radiation incident on a water surface penetrates to substantial depths. Water currents further distribute heat in water bodies and evaporation consumes heat energy at the surface. These various factors largely account for the fact that continents are colder than the oceans in winter and warmer in summer. In fact, ocean temperatures change slowly and by small amounts throughout the year. The oceans store tremendous quantities of heat and act as a thermostat for reducing global temperature changes.

Lab 10: Even You Emit Radiation

INTRODUCTION

Given the bad rap that radiation has received after Chernobyl and other historical events, it may be a bit disconcerting to find out that all objects emit radiation all the time, even you. It turns out that by measuring the emission from an object carefully and applying Stephan-Boltzman's Law, the surface temperature of the object can be determined from a distance. This approach can be used to gage a baby's fever by measuring the IR radiation with in their ear canal. It also provides an important tool to meteorologists for viewing the atmosphere and Earth. By viewing the emission from the ocean in areas free of clouds, the distribution of sea surface temperature can be estimated by satellite. Similarly, the temperature of cloud tops and thus their heights can be estimated from space.

As mentioned earlier, all objects absorb and emit radiation. To maintain a steady temperature, all objects, living and non-living, must emit the same amount of radiant energy that is absorbed. If this were not the case, the temperature of the object will either rise or fall depending on the nature of the imbalance. For instance, if you park your car in the sun in the summer very soon the dark dash will become quite hot. When you get in your car and drive, the dash will continue to give off a lot of infrared radiation for some time as it cools. This will cause your face to feel heat coming off the dash, despite having the air conditioning turned up.

ACTIVITY

OBJECTIVE: Introduce the basic concepts of radiative transfer and radiative balance. Also, to use a hand-held infrared pyrometer[1] to measure the radiation (temperature) of different objects.

---

[1] Hand held infrared pyrometers are available from numerous vendors at under $100.

MATERIALS:

§ Infrared pyrometer

§ Thermometer

§ Map of your school's campus

PROCEDURE:

1. Use the infrared pyrometer to obtain radiation temperatures for the following items (°F). Also make measurements of the air temperature (°F) above selected surfaces. Shade the thermometer with your body to avoid contamination from direct sunlight.

Shaded sidewalk: _____ Air temperature _____

Sunny sidewalk: _____ Air temperature _____

Shaded parking lot: _____ Air temperature _____

Sunny parking lot: _____ Air temperature _____

Grass: _____ Air temperature _____

Roof top _____ Air temperature _____

Cloud free area of sky: _____

Cumulus (if any): _____

Your hand: _____

Your head: _____

Your chest: _____

Roof of a white car _____

Roof of a black car _____

2. Use the observations made by the infrared pyrometer, shade the map of your school's campus, based on a color scale (enhancement curve) that you define for the range of temperature you measured.

QUESTIONS:

1. What law of radiation allows us to estimate the temperature of an object from the IR radiation it emits?

2. Which is more effective at absorbing infrared radiation, asphalt or cement?

3. Do you think your infrared map will look the same if you took your measurements at midnight? Why?

Lab 11: Distance from the Sun

INTRODUCTION

The fact that Earth is the only planet in the solar system that supports life is a direct consequence of its distance from the sun. If Earth were only two percent of its present distance farther away from the sun, it would be like Mars, a permanent "Ice Age" wasteland with a carbon dioxide atmosphere and all of its water tied up in polar ice caps. If Earth were only five percent closer to the sun, it would be like Venus, a planet many astronomers have described as a "hellish place". The surface temperature on Venus is about 850°F. Earth's distance from the sun is just right, and practically no other distance will do. Only recently, it has been determined that the range of distances from the sun in which Earth's conditions could have formed is very small compared to the scale of the solar system. Because of this narrow range (or "clement zone"), Earth's atmosphere is the only one in the solar system that will allow water to exist in all three states simultaneously- solid, liquid and gas.

This activity is designed to show how distance from a light source will affect temperature and that the range of distances in which a specific temperature can exist is relatively small.

ACTIVITY

OBJECTIVE: The objective of this activity is to investigate the relationship between distance from a light source and temperature.

MATERIALS:
§ ruler
§ 4 identical thermometers
§ reflector lamp
§ clay or tape
§ watch

PROCEDURE:

1. Place the ruler on a table and attach each thermometer to it with clay or tape at the correct distance (Table 3.1) to represent each planet (Fig. 3.1). Label each thermometer with the name of the planet it represents. Let one astronomical unit (AU, the mean distance between the sun and Earth) = 10 cm.

2. Adjust the lamp so that it is on the same plane as the thermometers (Fig. 3.1).

3. Record the starting temperatures for each thermometer in the Data Table.

4. Turn on the lamp. Observe and record the temperature of each thermometer every three minutes for fifteen minutes in the Data Table provided.

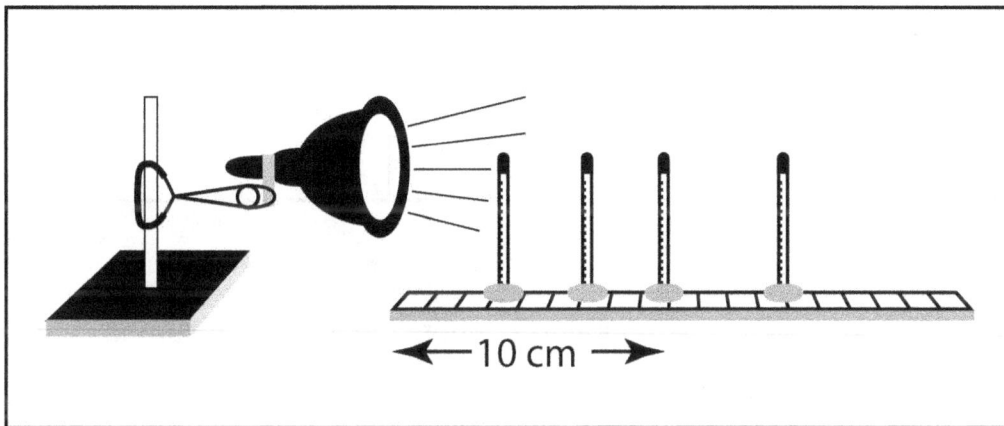

Figure 3.1 Schematic diagram

Table 3.1 Distance to the sun in astronomical units (AU) for select planets in our solar system

PLANET	DISTANCE FROM SUN (AU)
Mercury	0.38
Venus	0.72
Earth	1.00
Mars	1.52

Data Table

PLANET	Scale Distance	TEMPERATURE (°C)					
		Start	3 min.	6 min.	9 min.	12 min.	15 min.
Mercury							
Venus							
Earth							
Mars							

QUESTIONS:

1. What happened to the temperatures when the light was turned on? Did the thermometers heat up immediately?

2. Which thermometer showed the greatest rise in temperature? Least rise?

3. Make a graph of the temperature versus distance. Is the relationship linear (a straight line)? Explain the shape of the resulting curve.

4. Why is the hottest time of the day around 3:00 PM even though the sun is at its highest point in the sky at noon?

Lab 12: Surface Heating

INTRODUCTION

We've all walked across hot black top in a parking lot with our bare feet in the summer. Similarly, it is cooler to wear a white shirt in the hot sun than a black one. Why is this so and what affect does it have on the world around you?

The darker a surface is, the more light it absorbs and the faster it heats up. Conversely, a surface appears light when it has a relatively high reflectivity (albedo) for visible light. The relative ability of different surfaces to absorb light is referred to as absorptivity. The higher a surface's absorptivity the more light it absorbs and is converted to heat. A perfectly black surface absorbs 100 percent of the light that strikes it (albedo = 0). A perfectly white surface reflects 100 percent of the light that strikes it (albedo = 1).

The air above a warm surface is heated as the surface absorbs increasing amounts of light. As the air heats, it becomes less dense and rises. A convection current is generated as cooler air replaces the warmer air and then is subsequently heated as well. This convective motion contributes to the formation of winds.

When adjacent air over the ocean remains much cooler than that over the land, the warm air rises over the land and it is replaced by cooler air from over the ocean. The resulting circulation is referred to as a sea breeze. The temperature contrast from sea to land can be large due the fact that the ocean has a much larger specific heat than the land, and sunlight is absorbed through a greater depth of water. The opposite of a sea-breeze circulation often develops at night when the land cools more quickly than the ocean, and the resultant wind blowing off shore is called a *land breeze*.

ACTIVITY

OBJECTIVE: The objective of this activity is to investigate the phenomena of differential heating of surfaces.

MATERIALS:

§ reflector lamp with 100-Watt bulb

§ 1 black cup

§ 1 white cup

§ two insulated lids with slits

§ two thermometers

§ ruler or meter stick

§ sand

§ water

§ piece of glass or Plexiglas

PROCEDURE:

1. Slide the thermometers through the slits in the insulated lids so that the bulb of each will be about half way down in the cups.

2. Place the lids on the cups and put the cups side by side, about 10 cm from the lamp as shown in Fig. 3.2.

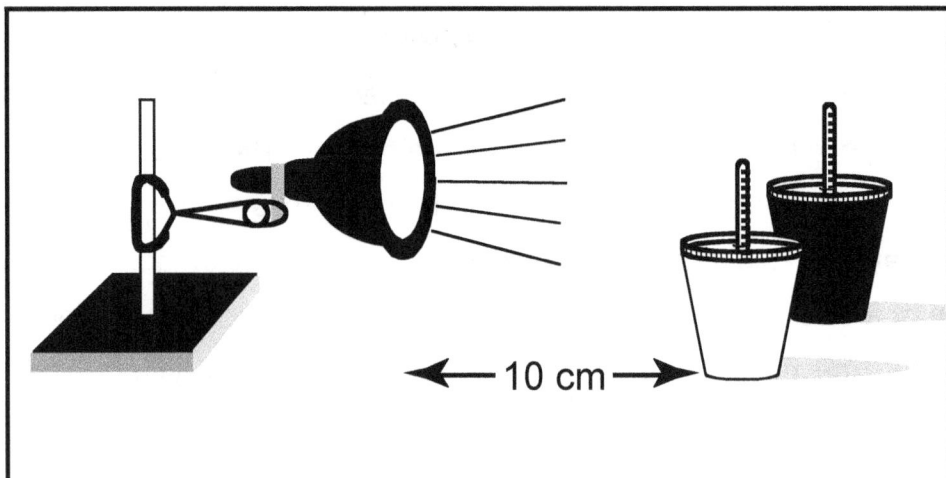

Figure 3.2 Schematic diagram

3. Record the initial temperature of each cup in the Data Table. In the space below predict what will happen to the temperature in each cup when the light is turned on, and explain your prediction. *Will there be any differences between the cups?*

4. Turn on the light and record the temperatures every two minutes for ten minutes.

5. Record your results in the Data Table below and graph your results on graph paper.

6. Place a piece of glass between the light source and the thermometers and repeat steps 3, 4, and 5. Is there any difference in the rate of temperature change with the glass inserted? Explain briefly below.

7. Fill one black cup with dry sand and the other black cup with wet sand. Repeat steps 1 through 5 above. In the space below predict what will happen to the temperature in each cup when the light is turned on, and explain your prediction. *Will there be any differences between the cups?*

DATA TABLES

	TEMPERATURE (° F) vs TIME (MINUTES)					
	0	2	4	6	8	10
BLACK						
WHITE						

	TEMPERATURE (° F) vs TIME (MINUTES)					
	0	2	4	6	8	10
BLACK						
WHITE						

	TEMPERATURE (° F) vs TIME (MINUTES)					
	0	2	4	6	8	10
DRY SAND						
WET SAND						

QUESTIONS:

1. What happened to the temperature measured in the two cups? Was your prediction correct?

2. How, if at all, would the results of this experiment have differed if a silver cup had been used instead of a white one?

3. What would happen if you were to leave the cups under the light for a long period of time, say 24 hours?

4. A glider is an airplane with no engine. In order to stay in the air, glider pilots sometimes look for fields that have been recently plowed. Explain why.

5. What is an advantage of spreading dark sand on snowy roads in locations where it snows a lot in the winter?

Lab 13: Solar Radiation

INTRODUCTION

Among the many consequences of the Earth being a sphere is the fact that the sun heats its surface differentially. The equatorial regions of the Earth receive the maximum amount of solar radiation, whereas polar regions receive a minimum. Snow and clouds enhance the impact of the unequal heating, by preferentially reflecting more of the available sunlight back to space at high latitudes. Any global view of the Earth will verify that this is the case (see the photo at the beginning of Chapter 7).

When rays of light strike the surface of an object at a 90° angle, the light is said to be *direct*. When the angle is anything less than 90°, the light is *slanted*. Sunlight strikes the Earth along a continuum from most direct at low latitudes to increasingly slanted at higher latitudes. At the equator, the Earth receives nearly direct sunlight and at the poles it receives the most slanted sunlight. As the angle between in the incoming sunlight and the Earth's surface increases, the area over which a unit of sunlight is spread also increases. Consequently the equator is always warmer than the poles.

If the Earth were not rotating, this differential heating would set up one global circulation cell with warm air at the equator rising and traveling toward the poles and cold air at the poles sinking and moving toward the equator. The Earth's rotation makes for a much more complex global circulation pattern that involves several convection cells (this point is returned to in Chapter 6).

ACTIVITY

OBJECTIVE: The purpose of this activity is to investigate the differential heating of the Earth's surface because of the combined affects of geometry and albedo.

MATERIALS:
§ Flashlight
§ 3 identical Celsius thermometers

§ reflector lamp with clamp and 60-watt bulb

§ ring stand with iron ring

§ utility clamp

§ black construction paper

§ stapler

§ books to prop up the thermometer

§ meter stick

§ scissors

PROCEDURE:

Part I

1. Rest a flashlight with a narrow well-defined beam on a desk or table such that the front of the flashlight is exactly 6 inches (12.5 cm) from a white wall and measure the diameter of the beam cast by the flashlight on the wall when the flashlight is pointed directly at the wall (90° angle).

2. Rotate the flashlight so that its beam makes a 45° angle with the wall. Make sure the distance between the flashlight and the wall remains exactly 6 inches. Use a ruler and a protractor to measure the angle of the beam. Again, measure the diameter the beam casts on the wall. Repeat this procedure at a 60° angle.

3. Make a graph showing the relationship between beam diameter and angle of incidence. Describe the relationship. What happens as the beam angle approaches 0°, shining parallel to the wall?

Part II

1. Use the black, medium grey, and white construction paper to cover the bulb of each thermometer as shown. Cut a strip of construction paper 5 cm x 10 cm. Fold the paper in half and staple as shown in the middle panel of Fig. 3.3. Insert the thermometer.

Figure 3.3 Schematic diagram

2. Prop the thermometers as shown in Fig. 3.4. The white thermometer should be vertical, the gray one slanted, and the black one horizontal.

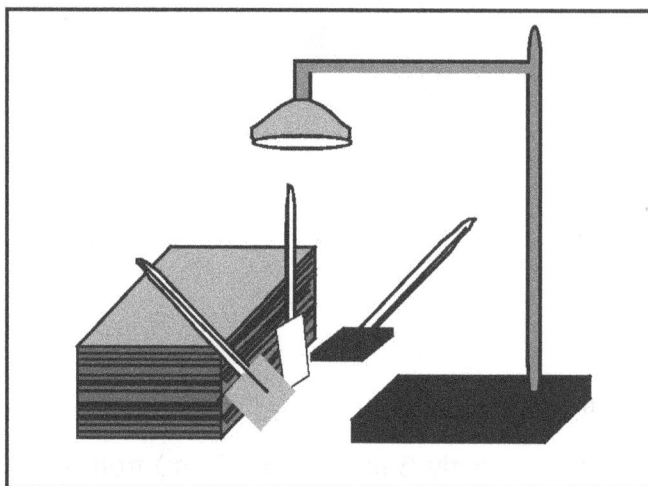

Figure 3.4 Schematic diagram

3. Adjust the lamp on the ring stand so that the bulb is centered 40 cm above the bulbs of the thermometers.
4. Record the temperature of all three thermometers in the Data Table under the zero column.
5. Turn on the lamp and record temperatures for each thermometer every minute for 15 minutes. Record all temperatures in the Data Table.
6. Using graph paper, make a graph of temperature versus time for each thermometer on the same graph. Using different colored pens or different types of lines (solid vs. dashed) will make the graph for each thermometer easier to distinguish.

Data Table

Time minutes	0	1	2	3	4	5	6	7	8	9	10	11	12	13	14	15	Total change
Vertical																	
Slanted																	
Horizontal																	

QUESTIONS:

1. Which thermometer showed the greatest temperature increase? Why?

2. Why did we use a different color construction paper for each thermometer?

3. If you were given a Data Table that listed the average yearly temperatures for cities as you go north from the equator, do you think you would see a trend in the temperatures? If so, what would this trend be and might there be exceptions to the general trend?

Lab 14: Local Climate

INTRODUCTION

There is a saying, "climate is what you expect and weather is what you get." This saying is accurate in so far as climate is the long term average of weather data. But, climate analysis also deals with extreme weather events. Recent years have brought wild swings in the weather, devastating storms, droughts, and floods. Many people presume these events are the product of a changing climate. But what is climate and what are the factors that control it? The temperature here today depends upon several factors -- seasons, latitude, altitude, proximity to oceans and prevailing weather patterns. In this exercise we will study a year's worth of temperature data and see what we can infer about the factors influencing climate at that location.

Climatological data are conveniently available for many locations in the form of a *Local Climatological Data (LCD) Annual Summary*. *LCDs* are several-page publications available in monthly and annual summary versions for close to 300 National Weather Service offices around the country. The detailed information they report includes temperature averages and extremes, precipitation amounts and extremes, wind values, skycover, snowfall, and heating and cooling degree days. Fig. 3.5 is taken from a *Local Climatological Data (LCD) Annual Summary* for Honolulu, Hawaii.

ACTIVITY

OBJECTIVES: The purpose of this activity is to investigate one way of comparing the short-term variations of weather and the long-term averages that help define climate. In particular, a year's worth of temperature values for a U.S. location will be compared with the longer-term climatic record at that location in order to

§ contrast differences between weather and climate data

§ use local climate data in planning seasonal activities

§ infer what factors influence climate

PROCEDURE:

To make this lab as relevant as possible to your location, obtain the *LCD* for a National Weather Service Observing Station near you. One can be obtained from your nearest National Weather Service office or by contacting the National Climatic Data Center, Federal Building, Asheville, NC 28801 (Telephone 704-CLIMATE; web URL http://www5.ncdc.noaa.gov/pubs/publications.html). Once you have obtained your local LCD, answer the questions that follow with respect to this LCD as well as with respect to the LCD provided in Fig. 3.5.

Daily temperatures for 1987 at the Honolulu International Airport National Weather Service observing station are plotted in Fig. 3.5. Each vertical line represents the actual daily temperature high, low, and range for that one day. The two curved lines that stretch across the graph describe normal high and low temperatures throughout the year. Normals are average values typically based on a recent thirty-year period. The upper solid curved line represents daily average maximum temperature, while the lower solid curved line represents minimum average temperature.

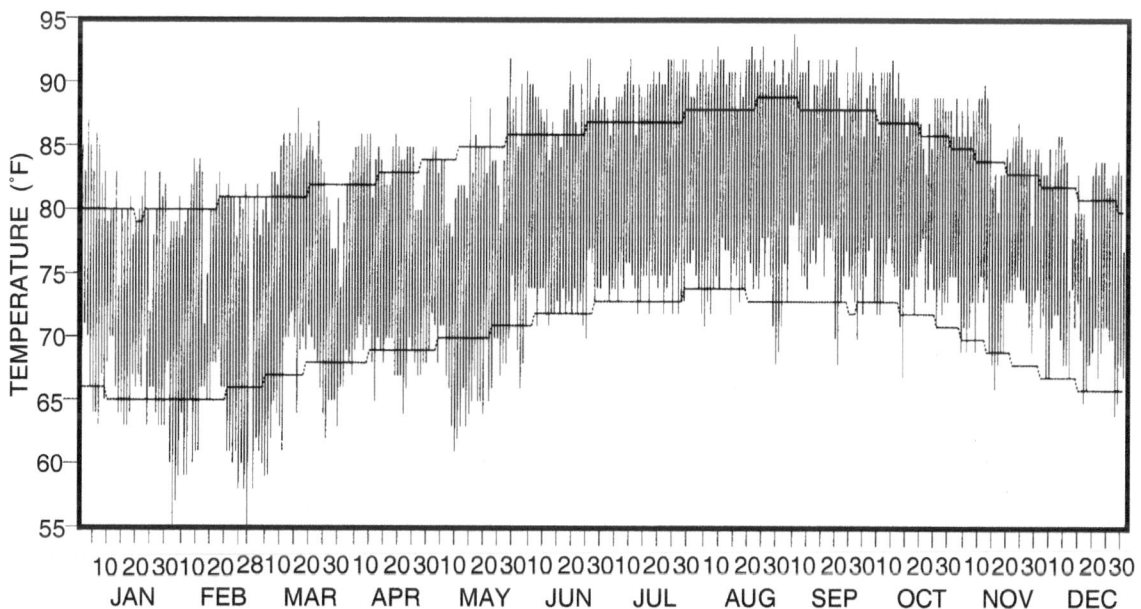

Figure 3.5 Annual Summary Of Daily Temperature Data From Honolulu, HI

QUESTIONS:

Refer to the Local Climate Data (LCD) Annual Summary for your location to answer the following questions.

1. In what month or months do the highest daily maximum temperatures *normally* occur at your particular location? When did the highest daily maximum temperature *actually* occur for the year given? What are their values?

2. The *average normal daily temperature* is the average of the *normal* high and low values for the particular day. Draw a curve on the graph to show *normal average daily temperatures* throughout the year.

3. People begin to turn on their air conditioners when the average daily temperature rises above 65 degrees F. For how many months would the air conditioning "season" last if temperatures were completely *normal* throughout the year at your location?

4. How does the daily temperature range of your location compare with that in Fig. 3.5? What factors influence this?

5. How does the seasonal temperature range of your location compare with that in Fig. 3.5? What factors influence this?

6. If temperatures were always *normal* at your location, when would the first fall frost be expected according to the LCD? According to the *actual* data reported in the graph on your LCD, when was the first date it could have occurred?

7. Referring to the Honolulu LCD in Fig. 3.5, would it surprise you to learn that few houses have air conditioning in Honolulu? What factors might reduce the need for air-conditioning in Hawaii?

Chapter 4 Weather's Invisible Fuel

'Water is the driver of Nature' Leonardo da Vinci

Living on a planet whose surface is three quarters ocean, water can easily be taken for granted. However, the contrast between a desert and a jungle illustrates the critical role of water to life on this planet. Novalis wrote 'our bodies are molded rivers,' alluding to the fact that we are more than 70 percent water. Water is equally essential to the circulations in the Earth's atmosphere, the formation of storms, and the Earth's climate of itself. It is the peculiar nature of water on the molecular scale that sets water apart from other substances, and clarifies the crucial role of water to the Earth's weather and climate and life. In this chapter some of the unusual properties of water will be explored, and ways in which invisible water vapor is measured in the atmosphere will also be discussed.

The water molecule is made up of one oxygen atom and two hydrogen atoms. Each of these atoms is made up of electrons with negative charge surrounding protons with positive charge. The geometrical distribution of these charges in the water molecule is such that on the side of the water molecule where the hydrogen atoms are attached there is slightly more positive charge than negative charge; conversely on the oxygen atom side of the water molecule there is slightly more negative charge (See Fig. 4.1). Because of this distribution, water is called a *polar* molecule. Therefore, when two water molecules approach, the positive side of one will be attracted to the negative side of the other. The resultant bond that forms between the two water molecules is called a *hydrogen bond*. When a hydrogen bond forms heat is released into the air, and conversely when a hydrogen bond breaks heat is consumed. Hydrogen bonds are relatively strong bonds, consequently the heat is associated with the bonds is large.

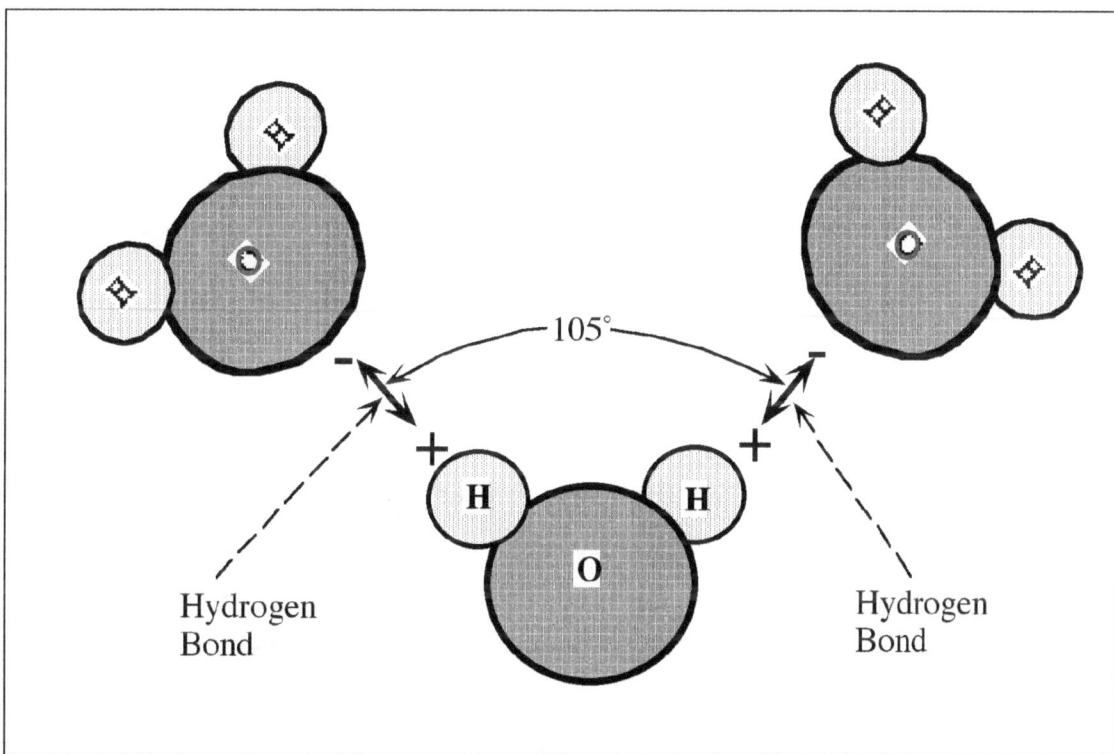

Figure 4.1 Bonding of water molecules

In liquid water there are millions of hydrogen bonds that rapidly break and reform, giving liquid water its fluid character. Consequently, liquid water is sometimes referred to as a *pseudo-crystalline* substance. It is the energy used in breaking hydrogen bonds that accounts for the large amount of heat required to heat a pan of water. Thus water is said to have a high *heat capacity*. Heat capacity is the ratio of heat absorbed by a substance to its corresponding rise in temperature. Specific heat is a measure of the heat capacity of a substance. (Recall that it requires one calorie to raise the temperature of water by one degree Celsius.) The oceans are, therefore, great moderators of the climate. In the solid or ice phase, water molecules are tightly bonded to each other in a regular crystal lattice. The geometry of the lattice gives rise to the beautiful hexagonal and dendritic snowflakes. In the vapor phase all of the hydrogen bonds are broken and the water molecules reside in a solitary gaseous phase.

There are several ways in which water molecules can change phase. In each case hydrogen bonds are either formed or broken, and heat is released or consumed. The heat associated with such changes of phase is referred to as *latent heat*, since it remains "hidden" until the phase change occurs. The magnitudes of latent heats vary with temperature, but at zero degrees Celsius, the latent heat of melting is 80 calories per gram, and for evaporation it is 580 calories per gram. These relationships are summarized below and in Fig. 4.2:

(evaporation) liquid to vapor – hydrogen bonds broken, 580-cal/g latent heat consumed

(condensation) vapor to liquid – hydrogen bonds formed, 580-cal/g latent heat released

(melting) ice to liquid   – hydrogen bonds broken, 80-cal/g latent heat consumed

(freezing) liquid to ice   – hydrogen bonds formed, 80-cal/g latent heat released

(sublimation) ice to vapor   – hydrogen bonds broken, 677-cal/g latent heat consumed

(deposition) vapor to ice   – hydrogen bonds formed, 677 cal/g latent heat released

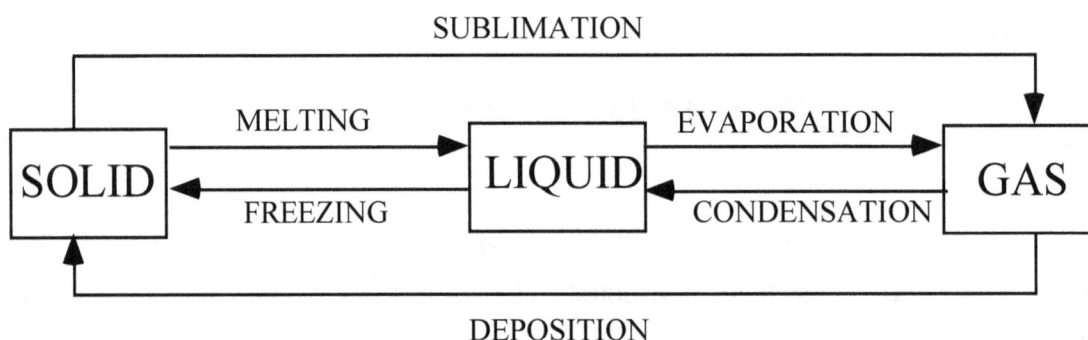

Figure 4.2 **Changes of phase in water**

Here are several common examples from daily life. When stepping from a shower into a dry room your wet skin feels cold. Heat is required to break all of the hydrogen bonds when water evaporates, and your skin provides much of that heat. Thus, until all the liquid water evaporates (or is removed by towel), your skin will continue to feel the cooling effect of evaporation. Conversely, when water droplets form on the outside of a glass of cold beer, condensation occurs and heat is released as the hydrogen bonds form, thus warming the beer. When clouds form, water molecules condense onto liquid cloud droplets, and heat is released into the cloudy air as the hydrogen bonds reform. This heat adds to the buoyancy of the rising cloud. It is the energy invested in the hydrogen bond that helps fuel thunderstorms, hurricanes and acts to moderate the climate of the Earth. More will be said about these topics in Chapters 6 and 8.

Measuring water vapor in the atmosphere

Having established the importance of water in transferring heat, how does one measure water vapor in the atmosphere? When air contains no water molecules, the vapor pressure is equal to zero; *vapor pressure* is the contribution to the total pressure by water vapor molecules. Given sufficient time, and a fixed temperature, the vapor pressure over a flat surface of water will reach equilibrium. In such a state, the rate at which water molecules leave the water surface equals the rate at which they enter the surface. At this point the air is said to be *saturated* and the *relative humidity* equals 100 percent. The relative humidity is the ratio of the actual vapor pressure to the vapor pressure at saturation, expressed in percent. In some circumstances, the quantity of water vapor in the air exceeds the saturation value, the air is *supersaturated*, and the relative humidity is greater than 100 percent.

For a given pressure, the saturation vapor pressure depends on the temperature and increases rapidly as the temperature increases. An illustration of the rapid increase in saturation vapor pressure with temperature is the fact that air at 40°C with relative humidity of 10% contains more water vapor than air at -10°C with a relative humidity of 100%.

Air over a surface of ice may also reach water-vapor equilibrium and become saturated with respect to ice. The saturation vapor pressure with respect to ice is less than the saturation pressure with respect to water because water molecules are more completely bonded to ice. Therefore, in a cloud composed of both ice crystals and liquid droplets, the ice will grow rapidly at the expense of the droplets. This point is important for the formation of precipitation and will be returned to in Chapter 5. An example of the growth of ice crystals at the expense of water drops is shown in the photo at the beginning of this chapter.

Early scientists noticed that certain natural fibers are sensitive to the amount of water vapor in the air. Instruments called *hygrometers* were constructed using animal hairs to measure the humidity of the air. Another way to measure water vapor is the *psychrometer*. A psychrometer employs two thermometers, one measuring the air temperature and the second measuring the wet-bulb temperature. The latter is obtained by covering a thermometer bulb with a wet muslin wick and ventilating it - as water evaporates, the temperature decreases. The lowest temperature attained is the *wet-bulb temperature*. Knowing the air temperature and the wet-bulb temperature, it is possible to determine from (psychrometric) tables the relative humidity or other measures of atmospheric humidity, such as the *dew-point temperature.* The dew point temperature is the temperature at which water droplets (or dew) form on a cooled, clean plate. This temperature is sometimes confused with the wet-bulb temperature. The wet-bulb temperature is actually a bit higher than the dew-point temperature because some of the water vapor from the wick moistens the air as the reading is taken.

Although many of the unique properties of water are discussed in this chapter several additional peculiarities of water have been deferred the chapter on clouds and precipitation (Chapter 5). Implications for the large latent heats associated with water's phase changes for the intensification of storms are discussed in Chapters 6 and 8.

Lab 15: Measuring Moisture in the Air

INTRODUCTION

Of all the basic molecules on Earth, one of the most important and most common is water. Water is the compound that has made the Earth the only life-bearing planet in the solar system. Much of the Earth's uniqueness may be attributed to the presence of water and its unusual properties.

The amount of water vapor the air can hold depends primarily on the temperature of the air. Warm air can hold much more water vapor than cold air. When warm humid air rises in thermals it expands, cools, and quickly becomes saturated. When this happens, water condenses out in the form of cloud droplets resulting in the formation of cumulus clouds so prevalent on warm summer afternoons. Similarly, when we set a cold drink out on a humid day, condensation quickly appears on the surface of the glass. The water condensing on the glass comes from the immediate surrounding air, which is cooled to its dew point.

Dew Point

One way to measure the water vapor content of the atmosphere is by the dew point temperature. The dew point temperature is the temperature at which water droplets first form on a cooled, clean surface. If water vapor does condense on a surface, this indicates that the temperature of the surface is below the dew point temperature. A higher dew point temperature indicates a higher moisture content; a lower dew point temperature indicates a lower moisture content. When the dew point temperature and the air temperature are the same, the relative humidity is 100%, and the air is said to be saturated.

ACTIVITY

OBJECTIVE: The purpose of this activity is to measure the amount of moisture in the atmosphere by observing the formation of dew.

MATERIALS:

§ water and ice cubes

§ thermometer

§ tin can with a shiny, clean surface

PROCEDURE:

1. Measure and record the temperature of the air on the data sheet.

2. Fill the can 3/4 full of water. Measure the temperature of the water as shown in Fig. 4.3, and record it on the data sheet.

3. Slowly add small pieces of ice to the can while carefully stirring constantly with the thermometer. Watch the outside of the can closely for the first sign of condensation. If condensation does not form before the ice cubes melt, add more ice cubes and continue stirring until it does form.

4. When the first condensation forms, immediately record the temperature of the water in the can under the column "Dew Point" in the Data Table.

5. Use a conversion table to convert the dew point temperature and temperature that you measured to relative humidity.

6. Repeat steps 1-5 outside.

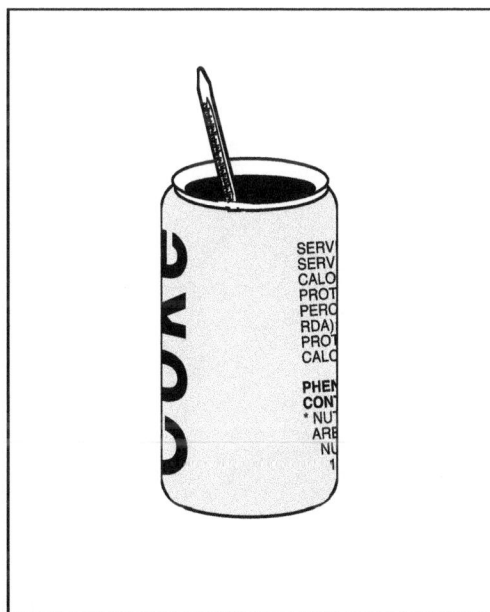

Figure 4.3 Schematic diagram

QUESTIONS:

1. When air rises, it expands and cools ~1 degree Celsius for every 100 m of altitude. If the air in the room were to rise, approximately at what height would the moisture begin to condense and form clouds? Show your work.

2. Repeat #2 for the air outside. Are there any clouds in the sky near this height?

3. What atmospheric conditions would allow water vapor from the air to condense on your skin?

4. How does the water vapor content of the air differ from inside to outside? What accounts for this difference?

DATA TABLES

Inside

Trial	Air Temp. (°C)	Dew Point Temp. (°C)	Relative Humidity
1			
2			
3			
Averages ⟶			

Outside

Trial	Air Temp. (°C)	Dew Point Temp. (°C)	Relative Humidity
1			
2			
3			
Averages ⟶			

Lab 16: Wet-Bulb Temperature

INTRODUCTION

We experience the cooling associated with the latent heat of evaporation in our daily lives. Each time we step out of the shower, evaporation provides the largest part of the chill we feel. More chill is experienced when the air in the room is dry than if the air is humid. Another common example is the cool feeling of grass on bare feet, also due to the evaporation of water from the grass blades. We are all familiar with the hot, humid summer afternoons that plague the East Coast. On these days, there is an abundance of water vapor in the air and our skin feels clammy and hot, because there is so little evaporation to provide cooling relief.

We can employ the cooling associated with evaporation to measure humidity in the atmosphere. This experiment will employ a sling psychrometer (Fig. 4.4) to measure the amount of moisture in the atmosphere by observing how far the latent heat of evaporation can depress the temperature of a "wet" thermometer.

ACTIVITY

OBJECTIVE: The purpose of this activity is to measure the moisture in the atmosphere through measuring the affect of the latent heat of evaporation on a wet thermometer.

MATERIALS:
§ water
§ sling psychrometer

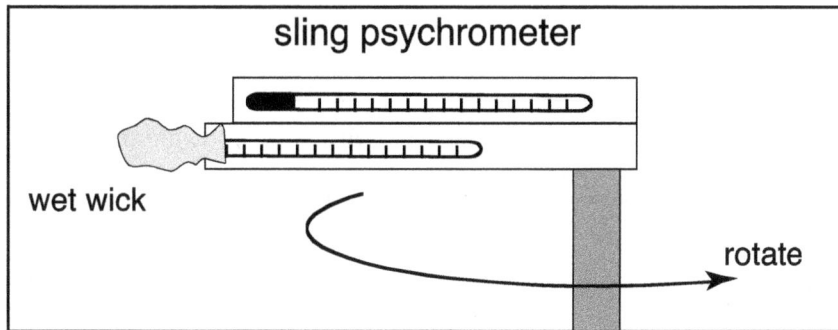

Figure 4.4 Schematic diagram of a sling psychrometer

PROCEDURE:

1. Using the sling psychrometer, measure and record the indoor air temperature (Fig. 4.4).

2. Wet the muslin of the psychrometer with water. Sling the psychrometer over your head for one minute. Make sure you have checked to make sure that nothing is in the way!

3. After a one-minute period, read the wet bulb temperature. Spin the psychrometer around for another thirty seconds. If the wet bulb temperature has not changed, record the temperature in the Data Table. If it has changed, sling the psychrometers for thirty-second periods until you get two readings that are the same. Record this temperature in the Data Table.

4. Repeat the above procedure outside.

5. Determine the relative humidity inside and out by referring to Table 4.1.

Data Table

Location	Time	Wet Bulb Temp. (˚C)	Dry Bulb Temp. (˚C)	Difference (˚C)	Relative Humidity
Inside					
Outside					

QUESTIONS:

1. Compare the relative humidity inside and out. What accounts for the difference?

2. Why do you sling the psychrometer until you have two similar readings?

3. If the temperature were to rise and the amount of water vapor were to remain constant, what would happen to the wet-bulb temperature and the relative humidity?

4. In the previous lab, the dew point temperature was measured. There is a difference between the dew point temperature and the wet bulb temperature. When measured at the same time under the same conditions, which of the two is higher? Why?

5 People living in hot and dry climates have long used porous unglazed pottery to store drinking water. Why?

6. On a global average, a layer of water about 100 centimeters in depth is evaporated to the air each year. About 600 calories of energy are required to evaporate one cubic centimeter of water at Earth-surface temperatures. Approximately how many

calories of energy are transported to the atmosphere above each square centimeter of the Earth's surface each year?

7. What happens to this energy when atmospheric water vapor condenses to form clouds?

8. Look at the global satellite view of the Earth at the beginning of Chapter 7 in this book. What does the distribution of clouds in the image suggest about how clouds act to redistribute heat over the globe?

Table 4.1 to Determine Relative Humidity

Dry Bulb Temperature (°C)

Difference Between Dry Bulb and Wet Bulb Temperature (°C)	5	6	7	8	9	10	11	12	13	14	15	16	17	18	19	20	21	22	23	24	25	26	27	28	29	30	31	32	33	34	35
1	86	86	87	87	88	88	89	89	90	90	90	90	90	91	91	91	92	92	92	92	92	92	92	93	93	93	93	93	93	93	94
2	72	73	74	75	76	77	78	78	79	79	80	81	81	82	82	83	83	83	84	84	84	85	85	85	86	86	86	86	87	87	87
3	58	60	62	63	64	66	67	68	69	70	71	71	72	73	74	74	75	76	76	77	77	78	78	78	79	79	80	80	80	81	81
4	45	48	50	51	53	55	56	58	59	60	61	63	64	65	65	66	67	68	69	69	70	71	71	72	72	73	73	74	74	75	75
5	33	35	38	40	42	44	46	48	50	51	53	54	55	57	58	59	60	61	62	62	63	64	65	65	66	67	67	68	68	69	69
6	20	24	26	29	32	34	36	39	41	42	44	46	47	49	50	51	53	54	55	56	57	58	58	59	60	61	61	62	63	63	64
7	7	11	15	19	22	24	27	29	32	34	36	38	40	41	43	44	46	47	48	49	50	51	52	53	54	55	56	57	57	58	59
8				8	12	15	18	21	23	26	27	30	32	34	36	37	39	40	42	43	44	46	47	48	49	50	51	51	52	53	54
9						6	9	12	15	18	20	23	25	27	29	31	32	34	36	37	39	40	41	42	43	44	45	46	47	48	49
10									7	10	13	15	18	20	22	24	26	28	30	31	33	34	36	37	38	39	40	41	42	43	44
11											6	8	11	14	16	18	20	22	24	26	28	29	31	32	33	35	36	37	38	39	40
12														7	10	12	14	17	19	20	22	24	26	27	28	30	31	32	33	35	36

Lab 17: Evaporation and Surface Area

INTRODUCTION

There is ample evidence in the open environment that water evaporates and enters the atmosphere. Surfaces dry after a rain. Water puddles gradually disappear. Clothes on a line dry in the open air. The question explored in this lab is how does the surface area exposed to the air effect the rate at which evaporation takes place and what are the practical implications?

ACTIVITY

OBJECTIVES: The objective of this activity is to observe the impact of surface area on the rate of evaporation and its implications in daily life.

MATERIAL:
§ one test tube
§ one glass beaker 50 ml
§ one Petri dish
§ about 15 ml of water
§ one heat lamp

PROCEDURE:
1. Pour 5 ml of water into each of the containers (test tube, beaker, Petri dish - or can substitute containers with similar dimensions and straight sides) and place them under the heat lamp. If no heat lamp is available, place the containers in a sunny location on a hot day.
2. Calculate the surface area of each container (area $A = 2\pi r^2$, where π = 3.14 and r = radius of the opening, See Fig. 4.5) and record the time required for the water to evaporate in the data table.
3. Make a graph of the evaporation rate (y-axis) versus the ratio of surface area to volume (x-axis).

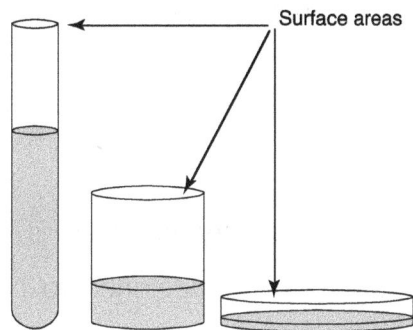

Figure 4.5 Schematic diagram

Data Table

	test tube	beaker	Petri dish
radius	cm	cm	cm
surface area	cm^2	cm^2	cm^2
ratio of surface area/volume	cm^2/ml	cm^2/ml	cm^2/ml
evaporation time	min	min	min
evaporation rate	g/min	g/min	g/min

QUESTIONS:

1. What happens to the evaporation rate as the ratio of surface area to volume increases? Explain with reference to your graph.

2. On the basis of your explanation in (1) how might an engineer choose the location for a dam in a dry region of the world?

3. On the basis of your explanation in (1) how might the size and shape of plants leaves change in going from a desert region to a rainforest region? Explain.

4. An economical air conditioner in dry climates is what is sometimes called a "swamp cooler". These devices are designed to cool air by evaporating water. Based on what you have observed in this activity, describe how they are best designed.

Lab 18: The Effect of Salt on Vapor Pressure

INTRODUCTION

When salt is dissolved in water, as it is in the ocean, the dissolved salt molecules residing on the ocean surface help prevent water molecules from evaporating. This causes the vapor pressure over sea water to be lower than it would be if the oceans were composed of fresh water. This fact reduces the availability of water vapor and the relative humidity in the atmosphere in contact with the ocean, and therefore, has implications for weather and climate. Since dissolved salt reduces the vapor pressure it also reduces the boiling point, the point at which the vapor pressure equals atmospheric pressure.

When an airborne salt particle forms the nucleus (cloud condensation nucleus) of a cloud droplet, the droplet that contains the dissolved salt will better resist evaporation than other droplets (Fig. 4.6). Therefore, salt particles from windblown ocean spray play an important role in the formation of marine clouds (more on this topic will be presented in Chapter 5).

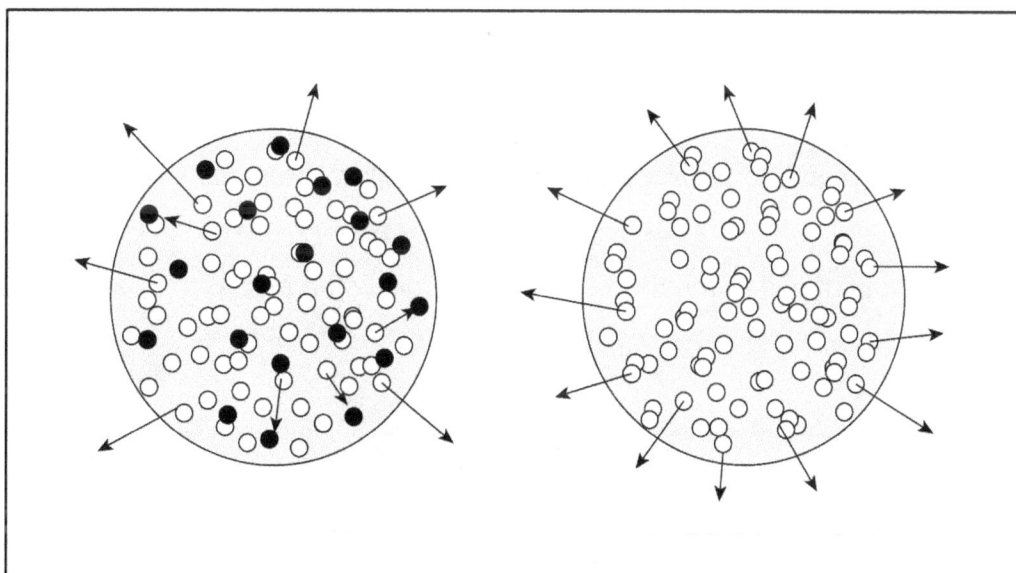

Figure 4.6 Schematic diagram showing the reduction in saturation vapor pressure caused by the presence of dissolved salt in a cloud droplet.

ACTIVITY

OBJECTIVES: The objective of this activity is to investigate the impact of the dissolved salt on the vapor pressure and the boiling point of water.

MATERIALS:

§ three 500 ml glass beakers

§ 300 ml water

§ salt

§ accurate thermometer

§ hot plate

PROCEDURE:

1. Fill each of the three beakers with 100 ml of fresh water.

2. Place the first beaker on the hot plate and bring the water to a boil. Record the time it takes for the water to come to a full boil in the data table. Carefully measure the temperature of boiling, to the nearest 1/10th °C if possible. Record this temperature in the data table.

3. Add 5 grams of salt to the second beaker and follow the steps in (2) above.

4. Add 10 grams of salt to the third beaker and follow the steps in (2) above.

Data Table

	Beaker 1	Beaker 2	Beaker 3
Water (solvent)	100 g	100 g	100 g
Salt (solute)	0 g	5 g	10 g
Boiling Point	°C	°C	°C
Extra time required to boil	s	s	s

QUESTIONS:

1. Based on the results in the data table, what is the relationship of salt concentration and vapor pressure (boiling point) in your experiments?

2. How can boiling points be used to determine the purity of liquids (e.g., mineral water versus distilled water)?

3. How does adding salt to a pot of vegetables effect the total cooking time needed for the vegetables?

4. If the oceans were fresh water instead of salty, how might this effect oceanic clouds?

Chapter 5 Clouds And Precipitation

'The most incomprehensible thing about the world is that it is comprehensible.'
Albert Einstein

Generally there is a balance between the upward-directed pressure-gradient force and the downward-directed force of gravity. This is called hydrostatic balance. When there is an imbalance between these two, vertical air motions arise. The difference between these two forces is called buoyancy or the buoyancy force. When a parcel of air is less dense than the surrounding air at the same altitude, the parcel is said to be buoyant and is subjected to an upward-directed force. At any height, the buoyancy depends mostly on the difference between the density of the parcel of air and the density of the surrounding environmental air. The density of the air depends mostly on temperature, but is decreased by the addition of water vapor and is increased by the addition of water drops or ice particles. For most purposes, the buoyancy force and upward acceleration of a parcel of air can be taken to be proportional to the difference between the parcel and the environmental temperatures.

The *stability* of the atmosphere is determined by the behavior of a parcel of air, which is displaced from one altitude to another and then released. If it continues accelerating in the direction of the displacement, the atmosphere is *unstable.* If the parcel returns to the original altitude, the atmosphere is *stable.* If the released parcel remains at the new altitude, the atmosphere is neutral.

When a parcel of air ascends in the atmosphere, it moves to a region of lower pressure and expands. In the process, its temperature decreases because energy is used to do the work of expansion. In the absence of any external heat sources or sinks (such as latent heat of condensation), the rate of decrease of temperature of an ascending or descending volume of air is 10°C per kilometer, and this is known as the *dry adiabatic lapse rate.* Rising air cools, while sinking air warms at the same rate, 10°C/km.

The term *lapse rate,* which is widely used in meteorology, is the rate of decrease of temperature with increasing height. Thus, the dry adiabatic lapse rate is +10°C/km. The stability of a cloudless atmosphere depends on the difference between the dry adiabatic lapse rate and the *environmental lapse rate.* The environmental lapse rate is the observed rate of change of the atmospheric temperature with increasing height. It normally is measured by means of a *radiosonde,* an instrumented package that is carried aloft by helium balloon. On average, through the lowest 10 km of the atmosphere, the environmental lapse rate is 6.5°C/km, but it varies widely from place to place and from time to time.

When the temperature increases with height, the lapse rate is negative, and a *temperature inversion* exists. In such a case, a rising parcel of air cooling at 10°C/km becomes increasingly cooler than the surrounding air. When this occurs, the atmosphere is considered to be stable because parcels return to their original levels.

Atmospheric stability is given according to the following differences between the dry adiabatic lapse rate (DALR) and the environmental lapse rate (ELR) (See Fig. 5.1):

ELR > DALR = unstable atmosphere

DALR > ELR = stable atmosphere

ELR = DARL = neutral atmosphere

When the atmosphere is unstable, there is enhanced vertical motion. One can visualize the upward and downward motion of air parcels or eddies. They transport such air properties as heat and pollutants from regions of high concentrations to regions of low concentrations. This process, known as eddy diffusion or turbulent diffusion, is important in mixing pollutants through the atmosphere. On a day when a temperature inversion is present, the atmosphere is stable, and there is little turbulent diffusion. As a result, pollutants released near the ground can become concentrated in a shallow layer. The behavior of plumes from smokestacks is governed to a large extent by atmospheric stability and the wind velocity.

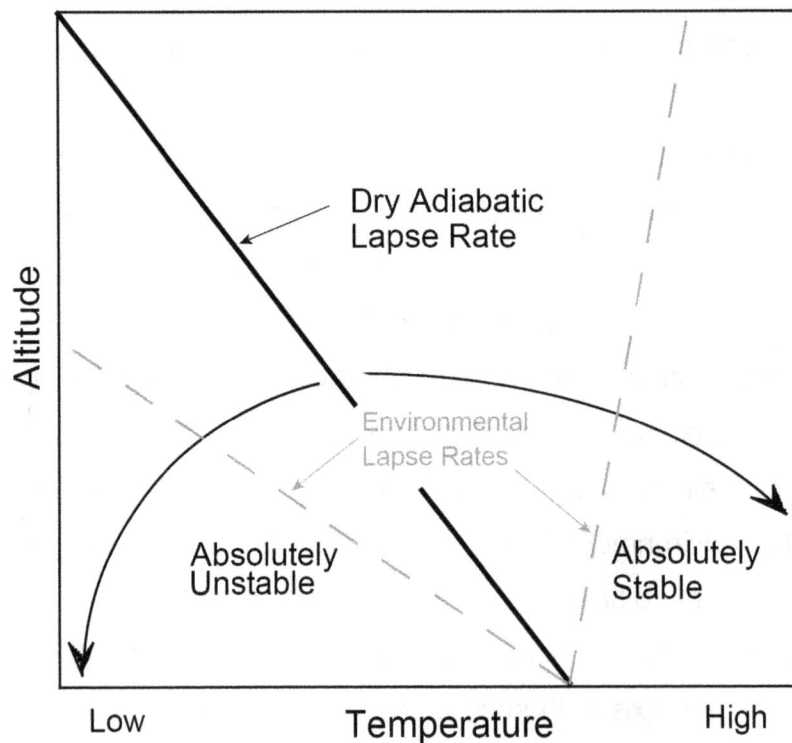

Figure 5.1 Stability diagram for dry air.

The nature of vertical motion can be greatly influenced by the condensation and evaporation of water in the atmosphere. When air is moist, that is, when its relative humidity is high, relatively small upward displacements lead to saturation, condensation, and cloud droplet formation. This process causes the release of latent heat of condensation. This heat partially compensates for the dry adiabatic temperature reduction resulting from expansion. The rate of decrease of temperature of an ascending air parcel within which condensation is occurring varies, depending on the amount of water condensing, but it averages about 6°C/km. This temperature change is called the moist-adiabatic lapse rate. A descending parcel of air warms at the moist adiabatic rate if the air contains water drops, which absorb heat as they evaporate while keeping the air saturated.

The latent heats released in the processes of condensation and freezing act to increase a cloud's temperature and increase its buoyancy. Thus the region of unstable lapse rates is greater under conditions of cloudy air than dry air (see *conditionally unstable* shaded area in Fig. 5.2). A process known as entrainment has the opposite effect. Entrainment is the mixing of cloudy and non-cloudy air across the cloud boundary. As the water and ice particles inside the cloud evaporate to saturate entrained dry air, the in-cloud temperatures are decreased. This reduces the cloud buoyancy. When clouds are relatively small and the surrounding air is very dry, entrainment can cause them to dissipate.

The water (hydrological) cycle

The total quantity of water in the Earth ecosystem is assumed to be unchanging. Nonetheless, there is a continuous exchange of water from the oceans and land to the atmosphere as a result of evaporation and transpiration. The water returns to the surface layers in the form of rain and snow. The sum of the water exchanges over the Earth is called the *hydrologic cycle.* On the average over the continents, precipitation exceeds evaporation. The excess water runs off into the oceans to compensate for the excess of evaporation over precipitation over the oceans.

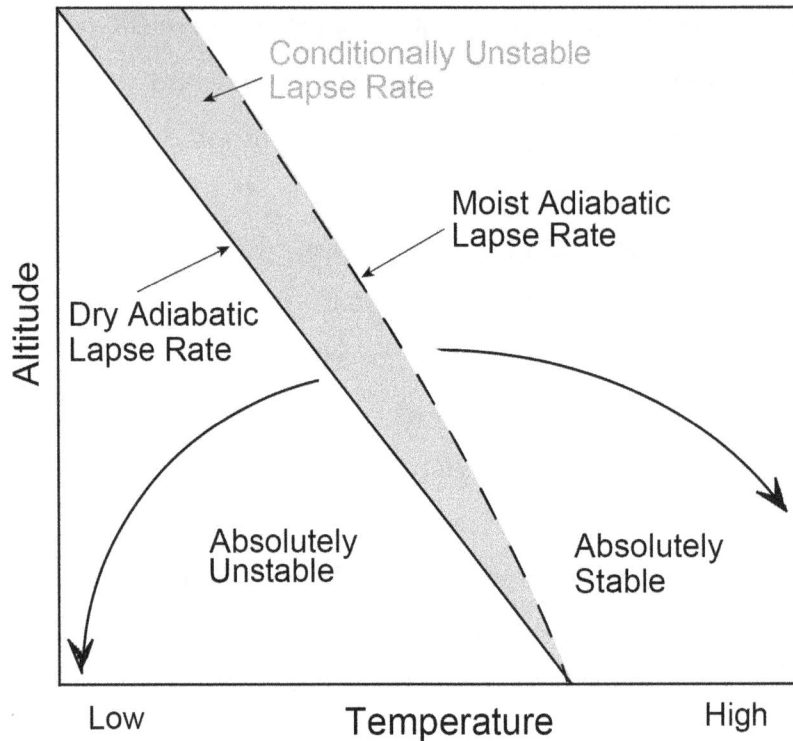

Figure 5.2 General stability diagram

For the Earth as a whole, average yearly precipitation amounts to a depth of water equal to 100 cm (0.27 cm/day). On the average, if all the water vapor in a column of air from the ground to the top of the atmosphere were condensed, it would produce a column of water 3 cm deep. If the atmospheric water vapor were rained out at the rate of 0.27 cm/day, it would take 11 days to "empty" the atmosphere of water vapor. This period is known as the average turnover or residence time of water vapor in the atmosphere. It is a statistically derived quantity which shows that water molecules, once evaporated into the air, tend to stay there a fairly short time and reflects the dynamic nature of the hydrological cycle. Even so, through the action of winds, water molecules are likely to leave the atmosphere as rain or snow at a point far from where they entered it.

On the average over the continents, precipitation exceeds evaporation. The excess water runs off into the oceans to compensate for the excess of evaporation over precipitation over the oceans.

Clouds

Clouds and precipitation are a dynamic part of the hydrological or water cycle on Earth. A cloud is composed of a very large number of small water droplets or ice crystals. The size and shape of the cloud depends on the characteristics of the atmosphere, particularly the moisture content, and the pattern of vertical air motions.

There are various ways to classify clouds. In the 1800's an English scientist named Luke Howard classified groups of clouds according to their appearances. He suggested 10 principal cloud types whose names were derived from Latin *(cirrus, cirrocumulus, cirrostratus, altocumulus, altostratus, nimbostratus, stratocumulus, stratus, cumulus, and cumulonimbus)* that fall in three major cloud subdivisions:

i) *cirrus* (hair-like) - high feathery clouds composed of ice crystals and usually occurring above ~6 km.

ii) *stratus* (layered) - clouds that form in layers

iii *cumulus* (pile or heap) - a detached cloud having the appearance of a mound, dome, or tower.

Some clouds have the characteristics of more than one of these categories. When a cloud is producing rain or snow, the term *nimbus* is added. For example, *nimbostratus* is a layer cloud producing precipitation. The prefix *alto* is added when clouds occur in the range from ~2 to 6 km in elevation.

After sunset and before sunrise in polar regions, beautiful clouds sometimes are seen at great altitudes that are brilliantly illuminated by the sun against the night sky. *Nacreous* clouds occur in the lower stratosphere at altitudes of 20 to 30 km, while *noctilucent* (literally "night light") clouds appear at altitudes close to 80 km.

Small cumulus clouds over continents are composed of water droplets mostly smaller than 20 μm (micrometers) in diameter that are present in concentrations of more than 1000 per cubic centimeter. In cumuli over tropical oceans, droplet concentrations may be as small as 50 per cm3, with the largest cloud droplets having diameters exceeding 50 μm. These differences are related to the greater abundance of particles in the air over continents than over oceans.

Tiny particles in the atmosphere, called *condensation nuclei,* are preferred sites for condensation to occur because they help reduce the retarding impact of surface tension on cloud droplet growth. They are introduced into the air from the ground and the oceans, as well as by combustion, both natural and manmade. Tiny particles may also be formed in the atmosphere from sulfur dioxide and nitrogen dioxide gases. Those particles, salty and acidic, on which condensation occurs readily are called hygroscopic nuclei. Condensation begins on such nuclei when the relative humidity of the air approaches saturation. Sea salt left in the atmosphere when ocean spray evaporates is an important source of hygroscopic condensation nuclei.

As water droplets are carried above the altitude where the temperature is 0°C, they generally do not freeze immediately. Instead, they remain liquid even at subfreezing temperatures and are said to be *supercooled.* Clouds often are supercooled to temperatures of -10°C to -15°C; on rare occasions, super cooling extends to temperatures as low as -40°C.

Ice crystals in the atmosphere occur at subfreezing temperatures above -40°C when *ice nuclei* are present. Ice nuclei are particles whose crystalline shape mimics that of ice. Certain soil particles are effective natural ice nuclei. Substances such as silver iodide, when they are dispensed as finely divided particles, can produce ice crystals at temperatures as high as -5°C. For this reason, they can be used to modify supercooled clouds.

Ice crystals develop in a number of shapes, each of which displays distinctive hexagonal (six-sided) characteristics. The most common shapes include: columnar—a long, narrow prism of hexagonal cross section; plate—a thin, solid plate having six sides; dendritic—a six-sided star, sometimes with each arm having an intricate, lacy structure.

Precipitation

Precipitation in the form of rain or snow occurs when particles of water or ice are large enough to reach the ground. Wisps or streaks of falling precipitation particles that evaporate before reaching the ground are called *virga.* The chief physical difference between a cloud element and a precipitation element is size. In terms of water drops, the boundary between cloud droplets and raindrops is generally assumed to be a diameter of ~200 μm. Water drops having diameters from 200 μm to

500 µm are sometimes called drizzle drops.

It takes many cloud droplets to make up a single raindrop. A raindrop 2 mm in diameter contains a million times more water than does a droplet having a diameter of 20 µm. In order for rain to occur, other drop growth processes, in addition to condensation, must occur.

Raindrops can be produced by the *collision* and *coalescence* of cloud droplets. Collisions take place because the terminal velocity of a water drop increases as its diameter increases, over the normal size range of cloud droplets and raindrops. Large droplets fall faster than, collide with and merge with smaller ones. When two rain droplets merge, coalescence has taken place. As a result of coalescence, the large drops can grow fairly rapidly.

Raindrops are also produced by the melting of ice crystals, snowflakes, and other frozen particles. When ice crystals exist in subfreezing air in the presence of supercooled water, the crystals grow as the droplets evaporate. This occurs because, at the same temperature, *the saturation vapor pressure over ice is less than that over water.* As a result, there is a pressure force driving the water molecules from the water to the ice, resulting in a rapid growth of ice crystals in the presence of liquid cloud droplets.

As ice crystals grow, the heavier ones fall. As a result, collisions and accretions occur. A snowflake is an aggregate of ice crystals stuck together. When such a particle falls through a layer of air whose temperature is above freezing, the crystals melt and raindrops are produced. In mountainous areas during the winter, valley locations often experience rain while snow falls at higher elevations.

Sometimes rain (melted snow) falls into subfreezing air beneath the front, the raindrops become supercooled, resulting in *freezing rain.* Freezing rain freezes when it makes contact with cold surfaces and produces hard, solid ice that is particularly hazardous to pedestrians and motor vehicles. If the cold layer is thick enough, supercooled raindrops will freeze in the air and reach the ground as small pellets of ice. Such precipitation is technically called *sleet.* Freezing rain and sleet occur most commonly in association with warm fronts in winter storms (see Chapter 6).

Lab 19: The Water Cycle

INTRODUCTION

The water (hydrologic) cycle is the movement and exchange of water substance among the atmosphere, ocean, and lands. Water vapor in the atmosphere plays a major role in the water cycle by returning water from land and ocean reservoirs across the globe, to even the highest elevations. Water vapor in the air condenses as it cools to form clouds, which return rain and snow to the Earth's surface.

Invisible to the eye, water vapor enters the atmosphere by evaporation. Evidence of this includes the often observed drying of sidewalks and roads after rain. Not so easily detected is the continual evaporation of water from soils directly, and through plants (transpiration). Once in the air, water vapor is carried aloft into the atmosphere through updrafts (or convection) and to other locations by the wind currents (advection). Ultimately, it changes phase to liquid or ice when the air is cooled to its dew point temperature. The following activity shows these processes on a small scale.

ACTIVITY

OBJECTIVES:

The purpose of this activity is to obtain evidence that water vapor enters the atmosphere from a variety of land surfaces and to describe, based on observations, differences in evaporation rates from various surfaces.

MATERIALS:

§ Clear plastic sheeting material

§ Black plastic sheeting material

§ Small weights

PROCEDURE:

Cut three one-meter square pieces of sheeting material. Place one over a grassy surface, one over bare soil, and one on an asphalt or concrete sidewalk (Fig. 5.3). Anchor the corners with weights to keep the sheets in place. After one half hour, observe the sheets of plastic and compare the relative amounts of condensation, if any, on their lower surfaces.

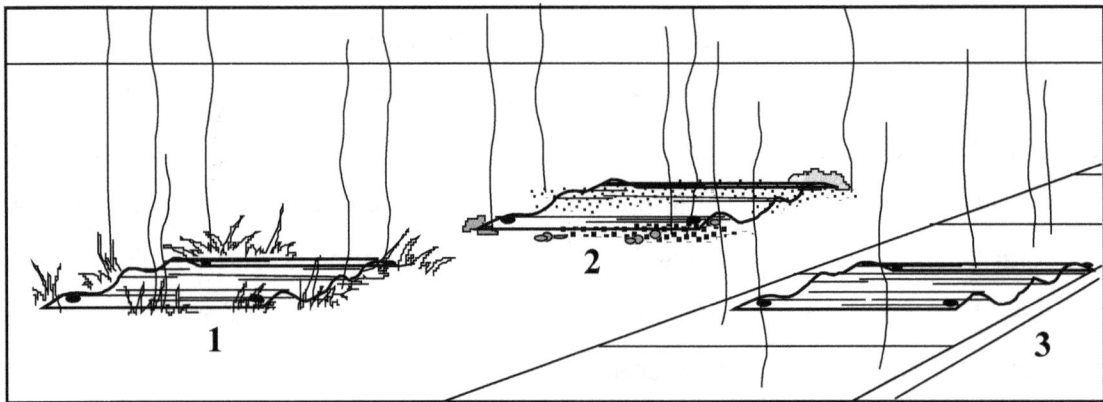

Figure 5.3 Schematic diagram

QUESTIONS:

1. Which land surface, if any, produced the most liquid water on the underside of the plastic sheets? The least?

2. Where did the water, if any, come from? What phase changes did it have to go through to end up deposited on the plastic sheets?

3. What can you conclude about different kinds of land surfaces and their relative influences in transferring water to the atmosphere?

4. How can your observations and conclusions be applied to describe the natural water (hydrologic) cycle?

5. Try this investigation with both clear and black plastic sheets. Do they produce different effects on the amount of evaporation? If so, why?

EXTRA CREDIT:

6. Try this investigation at various times of the day or under various cloud cover conditions. How does the brightness of sunshine appear to affect the rate of evaporation from the surfaces? Why?

7. Construct a broad, shallow "evaporation pan" and fill with water. Measure the drop in water level from day to day, and determine the rates of evaporation from the water surface under different weather conditions.

8. Design and conduct a version of this investigation to actually measure the rates of evaporation from different kinds of surfaces. One way might be to collect and weigh the water, which accumulates In an hour above each surface.

Lab 20: Recycled Water-The Hydrologic Cycle

INTRODUCTION

In considering the hydrology of the Earth, it is relevant to recognize that the oceans cover about 70% of the Earth's surface and that they represent more than 97% of the Earth's water. Another 2.2% is locked up in ice caps and glaciers. As global temperatures increase or decrease, the quantity of land-held ice decreases (by melting) or increases (by precipitation), resulting in important changes of sea level. The atmosphere contains only 0.001% of the planet's water, but this water crucially affects the lives of people, animals, and plants.

In the activity that accompanies this demonstration, a distillation apparatus is used to model the hydrologic cycle. It is important to keep in mind the distinctions between distillation and the hydrologic cycle as it occurs in nature.

ACTIVITY

OBJECTIVE: The objective of this activity is to investigate the hydrologic cycle.

MATERIALS:
§ clear plastic shoebox with lid
§ small plastic cup
§ Baggie filled with sand or soil
§ water
§ ice
§ lamp with reflector

PROCEDURE:
1. Set up the apparatus shown in Fig. 5.4.
2. Cut a hole in one corner of the lid of the shoebox, just large enough for the cup to fit halfway through the lid.

3. Add enough water to cover the bottom of the clear shoebox, a depth of about one inch.

4. Position the sandbag at one end of the box, directly under the opening of the cup.

5. Fill the small plastic cup with ice and place it in the opening of the lid.

6. Position a gooseneck lamp so that its light is shining onto the water inside the box.

7. Periodically check the set-up to observe the progression of events. Record your observation over the course of the class and diagram the movement of the water through the set-up.

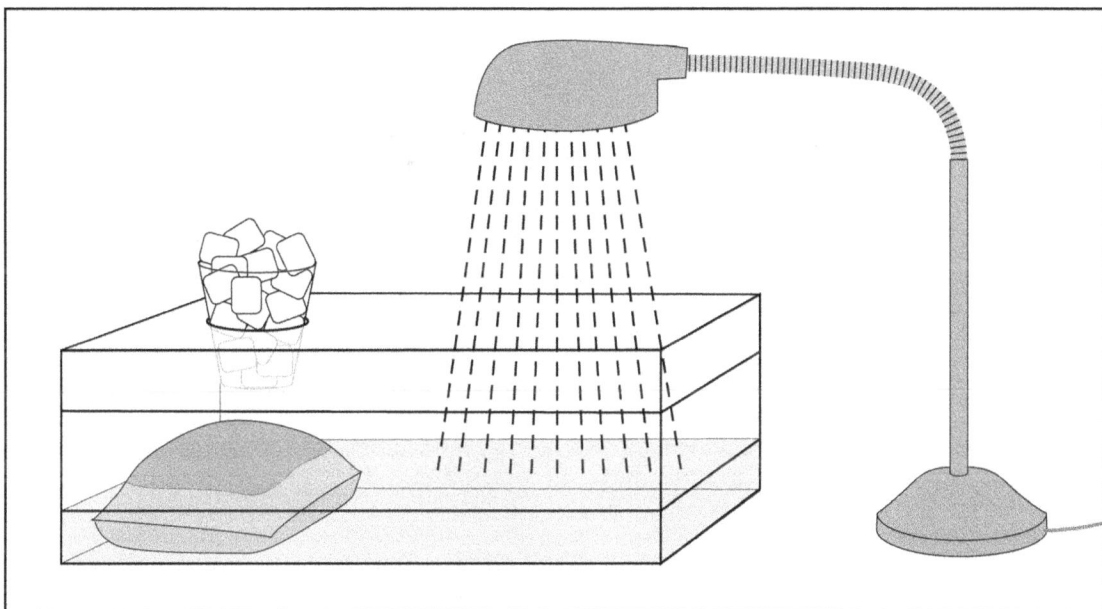

Figure 5.4 Schematic diagram

QUESTIONS:

1. Where in this activity does evaporation occur? Condensation?

2. What causes water droplets to form on the outside of the cup that is in the shoebox?

3. Where in the Earth's hydrologic cycle does evaporation occur? Condensation?

4. From what you observed in this activity, what are the key processes in the hydrologic cycle?

5. Was water lost from the system in any part of this activity? If so, where?

Lab 21: Formation of Clouds

INTRODUCTION

There are three conditions in the atmosphere that are met before a cloud forms. First, there must be sufficient moisture in the air. Secondly, the air must cool so that it becomes saturated and condensation can occur. And finally, there must be some type of particulate or condensation nuclei suspended in the air such as dust, smoke, or pollen for the excess water to condense on.

ACTIVITY

OBJECTIVE: The purpose of this experiment is to investigate the conditions, which must be present in order for clouds to form.

MATERIALS:
§ 32-oz. clear glass jar with lid
§ ice cubes or crushed ice
§ hot water
§ matches
§ can of aerosol spray (air freshener is best)
§ black construction paper
§ safety goggles
§ flashlight (optional)

PROCEDURE:
1. Fill the jar with hot water.
2. Pour out the hot water leaving only ~2 cm of water in the bottom of the jar. Place the jar in front of the upright construction paper.
3. Turn the lid of the jar upside down and fill it with ice. Place the lid on the jar as shown in Fig. 5.5. Observe the jar for three minutes. Darken the room and shine

the flashlight on the jar as you make your observations. Record your observations in the Data Table.

4. Pour the water out of the jar and repeat steps 1 and 2.

5. Move all loose papers away from the jar. Wearing your safety goggles, strike a match and drop the burning match into the jar.

6. Immediately cover the mouth of the jar with the lid full of ice as you did in step 3 and observe what happens for three minutes. Record your observations in the Data Table.

7. Pour out the water and repeat steps 1 and 2.

8. Spray a very small amount of aerosol in the jar and immediately cover the mouth of the jar with the ice filled lid. Observe what happens for three minutes and record the observations in the Data Table.

QUESTIONS:

1. Comment on the differences on your observations, and the reasons for these differences.

2. Suppose a layer of sand was put on the bottom of the jar instead of water, would a cloud form? Why or why not?

3. What would have happened in this experiment if cold water had been used instead of hot water? Why?

4. Describe any motion observed inside the jar. Explain your observation.

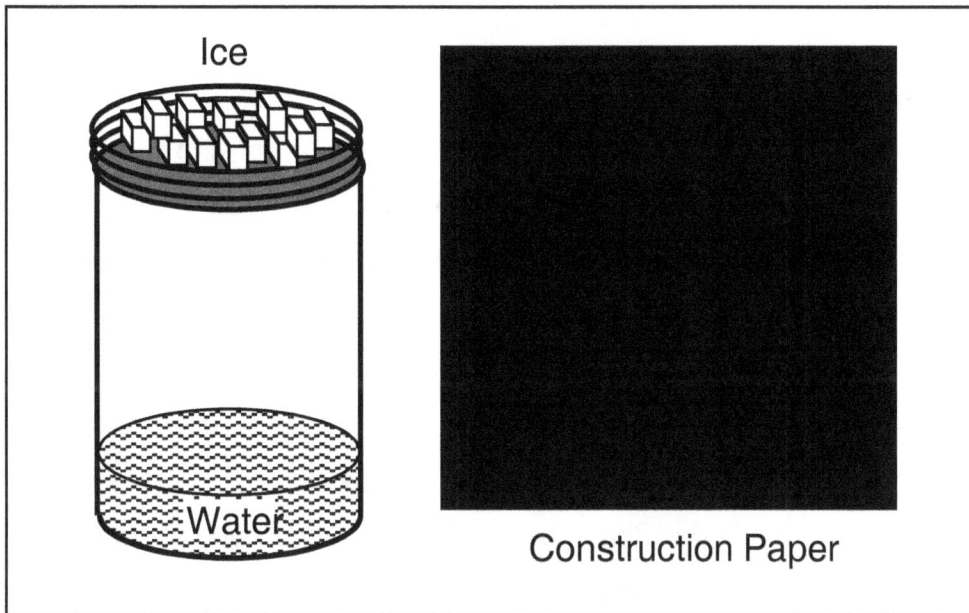

Figure 5.5 Schematic diagram

Data Table

TRIAL	OBSERVATIONS
No match or aerosol	
Match	
Aerosol	

Lab 22: Rain Makers

INTRODUCTION

Mountains have a profound impact on the weather, both on the wind field and on the climatology of cloud and precipitation. As moist air is forced upslope, it condenses and first forms clouds, and then rain, sometimes referred to as *orographic* clouds and rain. The direction of the flow at large scales interacts with the topography to produce gradients in cloud and rainfall, that in turn help define the local climate; e.g., wet windward and dry leeward slopes. The height of the mountains and the strength of the flow determine if the flow can pass over the top of the mountain or if the flow is force to split and go around the mountain. Mountains can act to both trigger and anchor thunderstorms, which can produce prodigious rainfall rates, up to 8 inches (200 mm) per hour. The fact that topographic features are fixed makes the interaction between large-scale storms systems and mountains somewhat easier to anticipate. The greatest accuracy in weather forecasting is associated with the largest scale systems, so that if one has detailed knowledge of topography, it is fairly straightforward to forecast the general character of the topographically forced clouds and rainfall that will occur during the passage of large-scale weather systems.

ACTIVITY

OBJECTIVE: The goal of this activity is to learn about the impact of mountainous terrain on the climatological distribution of wind and rainfall by analyzing wind and rainfall data collected during an extended field experiment called the Hawaiian Rainband Project (HaRP). You will also learn about orographic lifting and discover the relationship between rainfall amounts and islands terrain.

MATERIALS: Two maps are provided with contours for elevation plotted every 3000 ft and wind data plotted in knots per hour (Fig. 5.6), and rainfall data in mm (Fig. 5.7).

PROCEDURE:

1. On Fig. 5.6 draw streamlines. Streamlines are everywhere parallel to the instantaneous wind direction. Unlike contours, streamlines have no numerical value and can join other streamlines and even cross each other at col points in the wind field.

2. Label the leeward and windward sides of the island.

3. On Fig. 5.7 draw *isohyets* (contours of equal rainfall) every 20 mm (i.e. 20, 40,...160).

4. Label all your isohyets and label the areas of maximum and minimum rainfall with a large 'MAX' or 'MIN'.

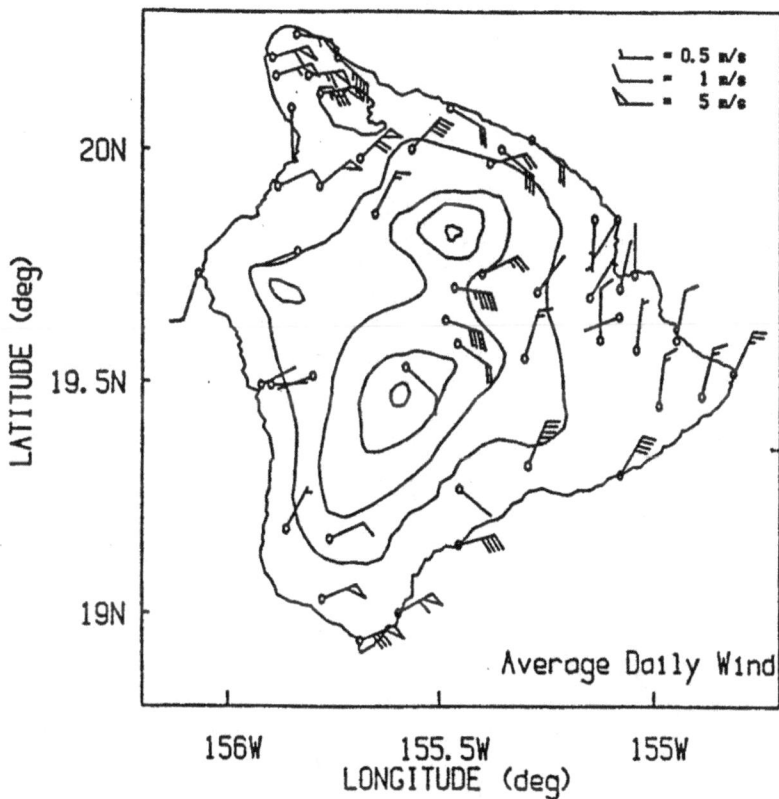

Figure 5.6 Hawaii Island with elevation contours every 3000 ft and average winds (kt) during a six-week period during July and August, 1990.

113

QUESTIONS (pertaining to Fig. 5.6):

1. Do your streamlines indicate flow splitting is taking place? If it is, explain why?

2. Where on average is the wind speed the greatest? Why?

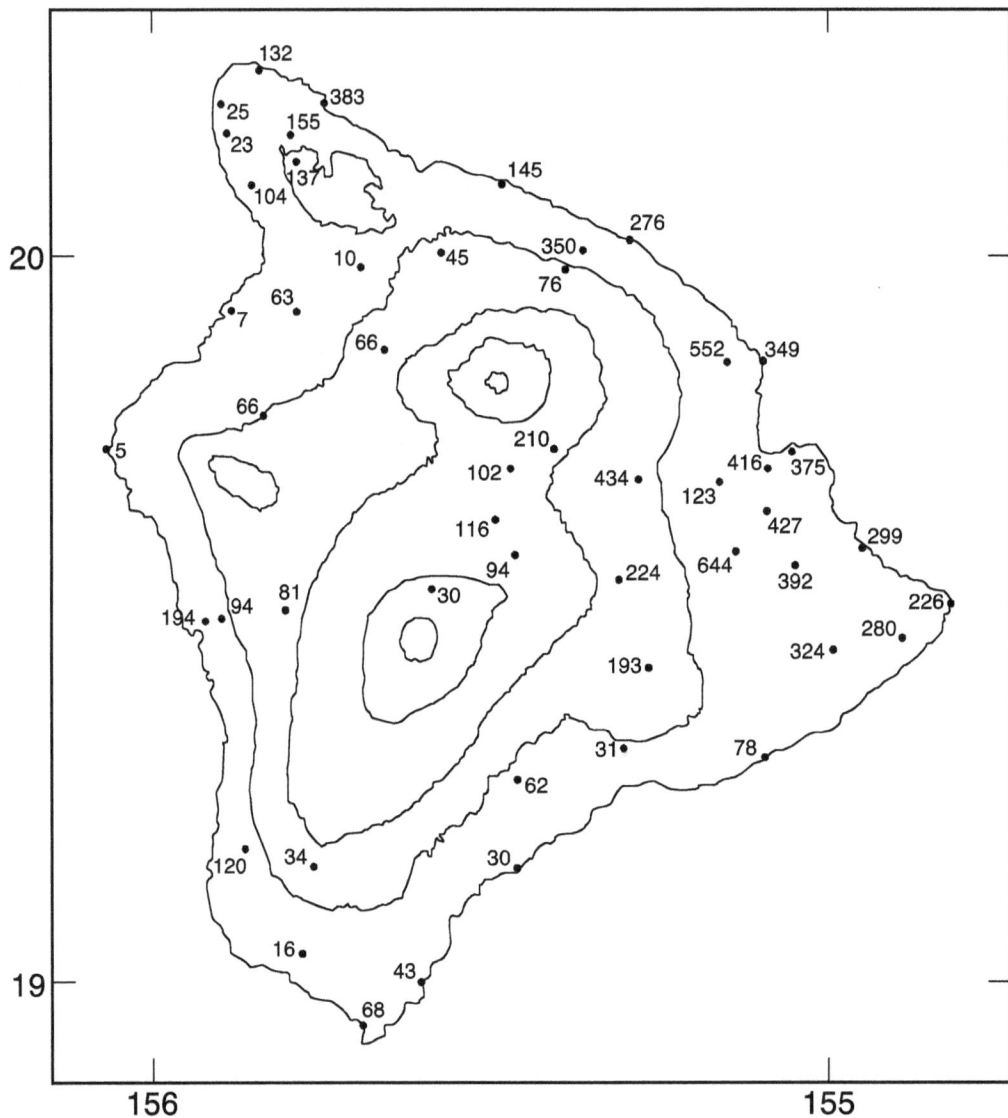

Figure 5.7 Hawaii Island with elevation contours every 3000 ft and average rainfall (mm) during a six-week period during July and August, 1990.

QUESTIONS (pertaining to Fig. 5.7):

1. Did you find the minimum rainfall where you expected to? What about the maximum rainfall? Explain!

2. Why are there no maxima in rainfall at the summits of Mauna Kea and Mauna Loa?

Lab 23: Archimedes' Eureka Moment

INTRODUCTION

Archimedes was a Sicilian scientist who was told by a king to find a way to determine whether or not a crown was made of pure gold. Some jewelers were not honest and would mix lead with gold when making crowns.

Archimedes was a hard working scientist. He often got so involved in solving problems and inventing things that he forgot to bathe. Then the king's soldiers would forcibly drag Archimedes off to the bath. According to the story, it was on one of these rare bath days that he saw, in a sudden epiphany, how to distinguish between a fake crown and one of pure gold, where upon he rushed naked into the street crying, "Eureka!"

Archimedes noticed that the water level rose as he lowered himself into the bath. This meant that his body displaced a volume of water. Since an object's density is equal to its mass per unit volume, there must be a relationship between that object's density and the density of a fluid in which it is immersed. From this, we get Archimedes' Principle: an object immersed in a fluid will experience a buoyant force equal to the weight of fluid displaced by that object.

The direction of the buoyant force is up, opposing the weight of the object. Furthermore, if the object floats (i.e., its density is less than or equal to that of the fluid), the magnitude of its weight will be equal to the magnitude of the buoyant force. By measuring the mass of the king's crown in air and in water and comparing those measurements, the crown's density could be determined. Today, we have more efficient methods for determining the authenticity of crowns, but Archimedes' Principle is still used to study the behavior of objects in fluids.

For example, we know that fish are naturally equipped to deal with the buoyant force. Many fish have a sac called a swim bladder that is filled with gas. By releasing or taking in gas, fish can control their overall density and avoid being forced to the surface or to the ocean floor. This knowledge of how fish control density has aided development of exploratory vessels used in underwater research.

ACTIVITY

OBJECTIVE: In the following activity, you will use Archimedes' Principle to predict what percentage of an iceberg is underwater. You will then make measurements to test the accuracy of your predictions.

MATERIALS:

§ 2 half-gallon milk cartons

§ a tank or large bowl

§ fresh water (from the sink)

§ salt water (50 grams of table salt to 1 liter of water)

§ measuring tape

PROCEDURE:

The day before doing this procedure, pour enough water to fill the bottom third of two half-gallon milk cartons and set them in a freezer to freeze.

The density of an object is its mass per unit volume. The density of ice is 0.92 g/cm3. The density of fresh water is 1.00 g/cm3. The density of salt water (35 ‰) is 1.03 g/cm3.

Since ice is less dense than fresh water or salt water, it will float.

The density equation is $\rho = m/V$ or $m = \rho V$; since $W = mg$, then $W = \rho Vg$, where

ρ = density   m = mass   V = volume

W = weight   g = gravity

1. Write an equation relating the buoyant force to the water's density, volume and acceleration due to gravity.

2. Write an equation relating the weight of an iceberg to its volume, density and acceleration due to gravity.

3. Using the two equations above, find the percentage of an iceberg that is below the surface of fresh water.

Repeat Step 3 for salt water.

Remove your icebergs from the milk cartons, and measure the volume.

4. Volume of iceberg in fresh water (Iceberg 1) = _____

5. Volume of iceberg in salt water (Iceberg 2) = _____

Very gently, place Iceberg 1 into a tank or large bowl of fresh water. As quickly and as accurately as possible, scratch out the line where Iceberg 1 breaks the surface of the water.

6. Measure the volume of Iceberg 1 that is above the surface of the water. (This part doesn't melt as much as the submerged part. What is the reason for this?) Subtract this number from the total volume of Iceberg 1 found in Step 4. Compute the percentage of Iceberg 1 that is below the surface.

Repeat the above procedure, placing Iceberg 2 in salt water.

Repeat the calculations in Step 6 for Iceberg 2.

7. Percentage below surface (fresh water) _____

8. Percentage below surface (salt water) _____

9. Compare your measured values with those you predicted earlier in #3.
   Use percent difference:

% difference = _____ = (measured value - predicted value) x 100

measured value

10. Finish the drawing of the iceberg shown below in Fig. 5.8. Sketch in the bottom of the iceberg to show how much ice is above and below the surface of the water.

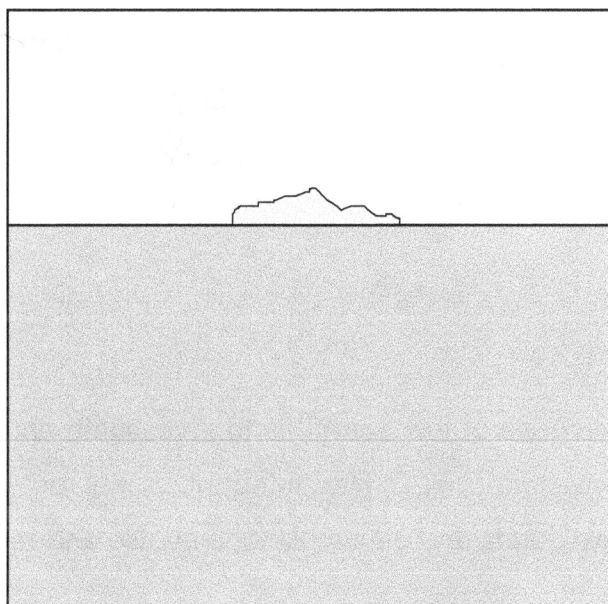

Figure 5.8 Schematic diagram

Lab 24: Convection

INTRODUCTION

Nearly all of the heat energy and moisture contained in the troposphere was input at the Earth's surface through a combination of conduction and turbulent mixing. This is because incoming sun light largely passes through the atmosphere unimpeded and is absorbed at the Earth's surface. Since the atmosphere is heated and moistened from below, convective currents redistribute this heat and moisture up through the troposphere. Stronger updrafts are visible as towering cumulus clouds and thunderstorms. Thunderstorm activity is especially prevalent in satellite imagery over land in the tropics (e.g., Amazon River Basin, Congo), where the solar radiation and surface heating are a maximum.

ACTIVITY

OBJECTIVE: The purpose of this activity is to investigate and observe how material moves within a convection cell. This information will be used to gain a better understanding of how clouds and air circulations evolve and redistribute energy in the Earth's atmosphere.

MATERIALS:
§ cafeteria tray
§ white paper
§ three Styrofoam cups
§ glass pan
§ water pitcher
§ food coloring
§ pipette or eyedropper

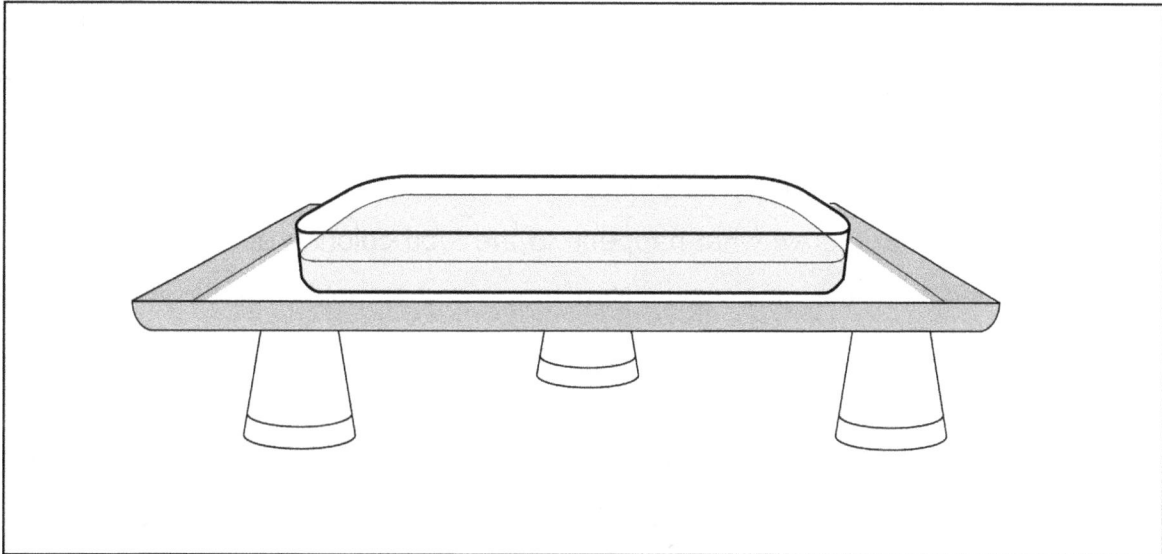

Figure 5.9 Schematic diagram

PROCEDURE:

1. Line the cafeteria tray with white paper to make it easier to observe the flowing food coloring.

2. Place three Styrofoam cups upside down on the paper. The fourth cup eventually will be placed right side up between the other three.

3. Fill the pan one-half to two-thirds full of water from one of the pitchers.

4. Place the pan on the inverted Styrofoam cups as shown in Fig. 5.9.

5. The water in the pan should sit for a minute or so before any food coloring is added to be sure there are no initial currents in the water.

TRIAL 1

Observe the movement of the food coloring when there is no heat source present. This will provide a basis for comparing the various trials, which do employ a heat source.

6. After the water in the pan has had an opportunity to sit undisturbed for a minute, place a very small drop of food coloring in the center of the pan. The food coloring

should be placed on the bottom of the pan. Move the pipette straight up and down in the pan of water and take care not to create currents in the water. Slowly release the drop of food coloring when the tip of the pipette touches the bottom of the pan, so as not to unduly disturb the circulation of the water.

7. On the data sheet, draw what happens to the food coloring as you look both from the top and from the side view. You may need to hold a piece of white paper behind the pan in order to see the food coloring better when viewing from the side. Write a brief description of what you see in the space provided on the data sheet.

8. Once you are satisfied that you have a clear understanding of the movement of the food coloring in this trial, gently swirl the water to disperse the food coloring. The water in the pan need not be changed after the first trial, unless the water becomes too dark to be able to observe movement of additional food coloring placed in the pan.

TRIAL 2

Observe the movement of food coloring when a heat source is placed directly beneath the center of the pan. Again, place the food coloring on the bottom, in the center of the pan.

9. Get a Styrofoam cup of hot water. Take care to avoid spilling the hot water. Carefully slide the cup underneath the center of the pan of water, and allow the water in the pan to become still.

10. Follow the steps as in Trial 1. After completing Trial 2 replace the water in the glass pan with cool clear water and refill the cup with hot water. Describe the results in the Data Table.

TRIAL 3

Observe the movement of food coloring with the heat source placed under the center of the pan and with the food coloring placed on the bottom and to one side of the center of the pan.

11. Follow each of the steps outlined in Trial 2, but for this trial, place a small drop of food coloring at the bottom of the pan about halfway between the center of the pan and the side. Remember to slowly release the food coloring when the tip of the pipette touches the bottom of the pan so as not to unduly disturb the circulation of the water. Describe the results in the Data Table.

TRIAL 4

Observe the movement of food coloring with the heat source placed under the center of the pan and with the food coloring placed to one side of the center of the pan on the top of the water.

12. Again, follow each of the steps outlined in Trial 2, but for this trial, place a small drop of food coloring about halfway between the center of the pan and the side. Place the drop of food coloring directly on the top of the water rather than on the bottom of the pan. Describe the results in the Data Table.

Data Table

	TOP	SIDE VIEW
TRIAL 1 heat = none color = bottom, center		
TRIAL 2 heat = center color = bottom, center		

Data Table

	TOP	SIDE VIEW
TRIAL 3 heat = center color = bottom, side		
TRIAL 4 heat = center color = top, side		

QUESTIONS:

1. Summarize the results in the data tables from the four trials by describing a convection cell in general terms. Draw a schematic model of a convection cell, showing the heat source and fluid motions.

2. If the water in the glass pan represents the atmosphere, what does the hot water in the Styrofoam cup represent?

3. The currents of water cause the food coloring to move at a rate of 2 to 3 cm or more per minute. By observing the growth of cumulus clouds on a fair day, how quickly would you say the air circulates through the clouds? How do these two figures compare?

4. Convection is a process that occurs not only in the atmosphere, but is also the driving force for other phenomena. Relate some other areas in which convection is important.

Lab 25: *Weather Balloons and Radiosondes*

INTRODUCTION

For over 60 years, upper air observations have been made by the National Weather Service (NWS) with radiosondes. The radiosonde is a small, expendable instrument package that is suspended below a polyurethane balloon filled with hydrogen or helium at the surface (Fig. 5.10). As the radiosonde is carried aloft, sensors on the radiosonde measure profiles of pressure, temperature, and relative humidity. These sensors are linked to a battery powered, 300-milliwatt radio transmitter that sends the sensor measurements to a sensitive ground receiver on a radio frequency ranging from 1668.4 – 1700.0 MHz. By tracking the position of the radiosonde in flight, information on wind speed and direction aloft is also obtained. Observations where winds aloft are also obtained are called "rawinsonde" observations. The radiosonde flight can last in excess of two hours, and during this time the radiosonde can ascend to over 35 km (about 115,000 feet) and drift more than 200 km (about 125 miles) from the release point. During the flight, the radiosonde is exposed to temperatures as cold as −90°C (-130°F) and an air pressure only few thousandths of what is found on the Earth's surface. When the balloon has expanded beyond its elastic limit and bursts (about 6 m or 20 feet in diameter), a small parachute slows the decent of the radiosonde, minimizing the danger to lives and property. Only about 20 percent of the approximately 75,000 radiosondes released by the NWS each year are found and returned to the NWS for reconditioning. These rebuilt radiosondes are used again, saving the NWS the cost of a new instrument. If you find a radiosonde, follow the mailing instructions printed on the side of the instrument. Although all the data from the flight are used, data from the surface to the 400-mb pressure level (about 7 km or 23,000 feet) are considered minimally acceptable for NWS operations. Thus, a flight may be deemed a failure and a second radiosonde is released if the balloon bursts before reaching the 400-mb pressure level or if more than 6 minutes of pressure and/or temperature data between the surface and 400 mb are missing. Worldwide, there are nearly a 900 upper-air observation stations. Most are located in the Northern Hemisphere and all observations are usually taken at the same time each day (00:00 and/or 12:00 UTC), 365 days a year. Observations are made by the NWS at 92 stations – 69 in the conterminous United States, 13 in Alaska, 9 in the Pacific, and 1 in Puerto Rico. NWS supports the operation of 10 other stations in the Caribbean. Through international agreements data are exchanged between countries.

How are radiosonde data used? Understanding and accurately predicting the evolution of weather systems requires adequate observations of the upper atmosphere. Radiosonde observations are the primary source of upper-air data and will remain so into the foreseeable future. Radio observations are applied to a broad spectrum of efforts. Data applications include, input in computer-based weather prediction models, analysis of stability for local severe storms, aviation, and marine forecasts, weather and climate change research, input for air pollution models and ground truth for satellite data.

Isothermal Layers: A layer in the atmosphere where the temperature remains constant. On the skew-t diagram the temperature remains constant (parallel to isotherms) throughout this layer. Isothermal layers are stable and resist vertical motion.

Inversion: A layer in the atmosphere where the temperature increases with height. Inversions are very stable, providing a *lid* to vertical motions. Sinking air in prevailing regions of subtropical high pressure produces an inversion that caps the trade wind flow, hence the name trade-wind inversion.

Lifting Condensation Level (LCL): The height where the air in a rising parcel reaches saturation and water vapor begins to condense into liquid water droplets (cloud) is called the LCL. From the environmental temperature at the surface follow the dry adiabat upwards. From the environment dew-point temperature follow the mixing ratio line upwards. The point where these two lines cross is called the LCL and coincides with cloud base.

Level of Free Convection (LFC) and Equilibrium Level (EL): After crossing the LCL, the parcel becomes saturated. Since it is saturated the parcel's temperature now follows parallel to a moist adiabat, because latent heat is released when water vapor is condensed to liquid water the lapse rate increases on average to –6 °C/km (dashed lines that angle leftward in Fig. 5.11). The level at which the parcel reaches the same temperature as the environment is called the LFC. At this point the parcel will rise on its own because it has become positively buoyant. The parcel will remain positively buoyant to the EL where the parcel temperature once again crosses the environmental temperature and becomes cooler and negatively buoyant.

Positive Buoyancy and CAPE : Areas where the parcel is warmer than the environmental temperature are positively buoyant and areas where the parcel is colder than the environmental temperature are negatively buoyant. The total area with positive buoyancy is called Convective Available Potential Energy or CAPE and is directly related to the magnitude of the upward motion that can be attained by the parcel.

Figure 5.10 Release of a radiosonde from the roof of the Atmospheric Sciences Department at the University of Hawaii at Manoa

ACTIVITY

OBJECTIVE: The objective of this activity is to investigate the thermodynamic information provided by radiosondes.

MATERIALS: Global radiosonde data from recent and historical weather balloons released by the weather services of countries around the world are available at http://weather.uwyo.edu/upperair/sounding.html. The radiosonde data can be viewed in various graphical formats including a skew-T diagram (Fig. 5.11) or they can be retrieved as a text file (Table 5.1). A blank skew-T diagram is provided in Fig. 5.12. However, larger color versions can be obtained from Purdue University Printing at: http://www.adpc.purdue.edu/PhysFac/prnt/Support/diagram.htm .

ACTIVITY: Given the plotted radiosonde data in Fig. 5.11 below answer the following questions.

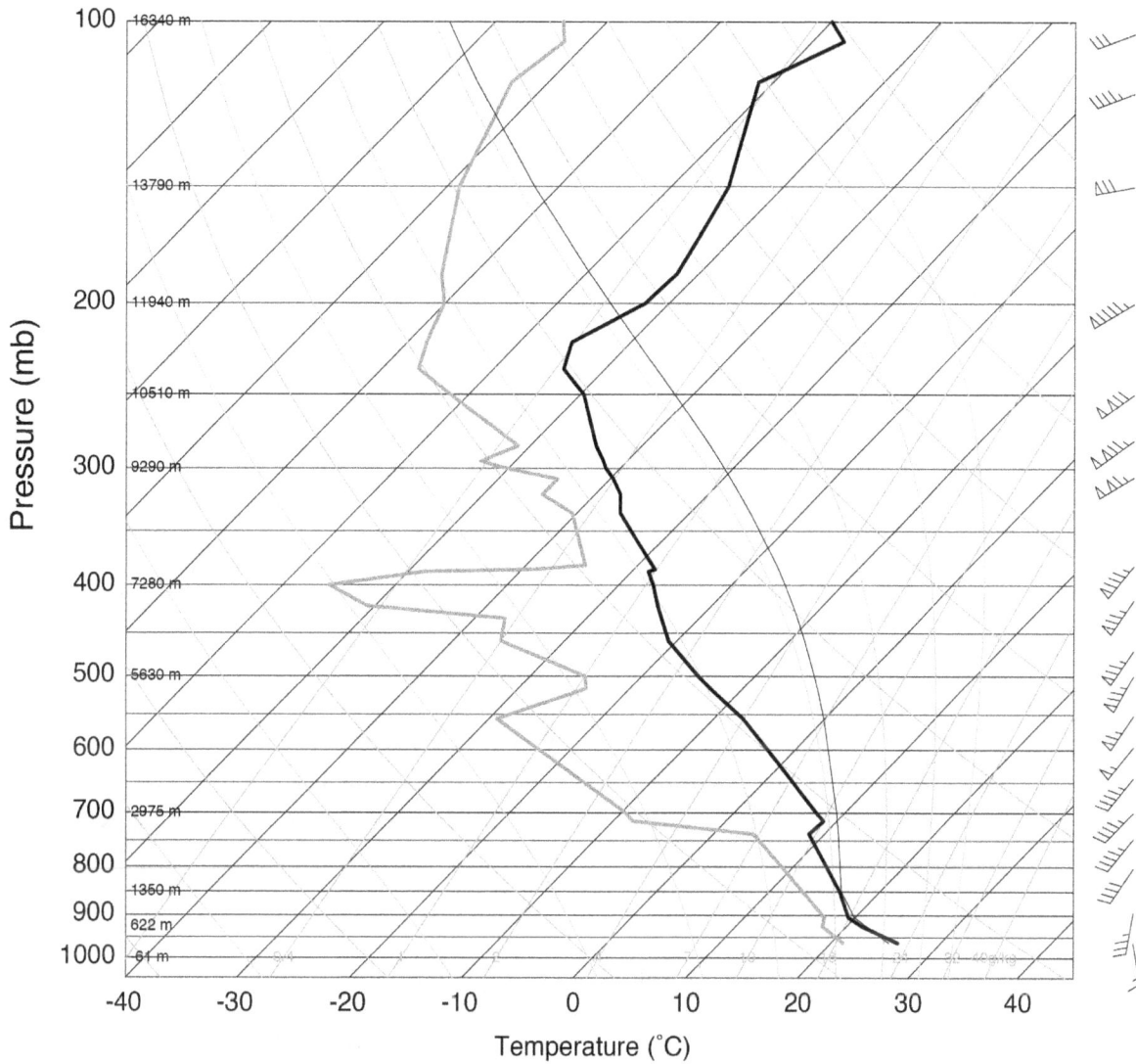

Figure 5.11 Skew-T diagram showing temperature (˚C) and dew-point temperature (˚C) as a function of height (pressure level inmb) for Topeka, KS at 1800 UTC on 4 May 2003.

QUESTIONS

1) Lift a parcel of air from the ground to 100 mb on Fig. 5.11. Lift it along a dry adiabat until the line intersects the mixing ratio line, then lift the parcel along a moist adiabat. Use a bright color so your work is easy to see.

2) Is this sounding stable or unstable? Why?

3) What is the level of free convection? (Give your answer in terms of pressure). What is the level of the tropopause, how did you identify it?

ACTIVITY PART TWO:

1) Plot the temperature and dewpoint temperature data provided in Table 5.1 on the skew-T diagram in Fig. 5.12 on the next page or on one provided by your instructor.

2) On the skew-T diagram identify and label the LCL.

3) From looking at the plotted data, identify where the trade wind inversion is. Is a thunderstorm likely to develop with the presence of an inversion? Why or why not?

Table 5.1 Topeka Radiosonde Observations for 18 UTC on 04 May 2003. Listed are Pressure, Height, Temperature, Dew Point Temperature, and Relative Humidity.

PRESS hPa	HGHT m	TEMP °C	DWPT °C	RELH %
--------	-------	------	-------	-------
1000	129	20.6	18.6	88
945.9	610	19.1	15.2	78
880.8	1219	14.7	12.5	87
798	2050	9.6	8.7	94
734	2742	6.6	5.3	91
700	3131	7.4	-24.6	8
656.5	3658	7.3	-26.8	7
609.4	4267	6	-27.4	7
565	4877	1.1	-28.9	9
500	5850	-5.1	-35.1	7
447.9	6706	-9.5	-37.6	8
400	7570	-16.7	-40.7	11
350.3	8534	-24.5	-45.8	12
300	9660	-33.7	-51.7	15
250	10910	-42.7	-57.7	18
200	12380	-54.5	-66.5	21
150	14160	-68.9	-78.9	22
100	16530	-74.5	-83.5	24

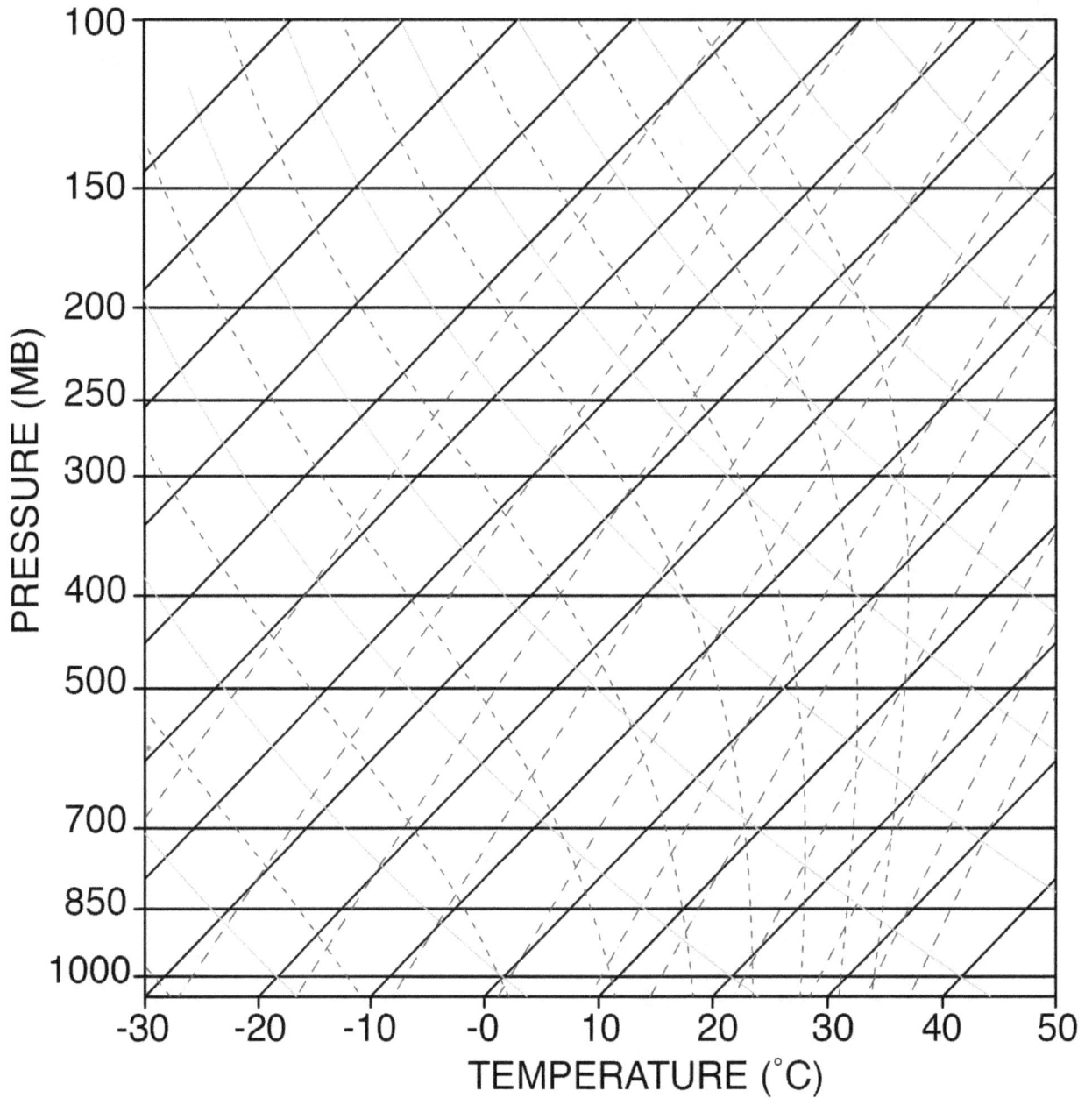

Figure 5.12 **Skew-T diagram with lines indicating constant pressure (solid horizontal), temperature (solid black angling right) dry adiabats (thin solid angling left), moist adiabats (dashed lines curving leftward), mixing ratio (dashed angling right).**

Rooftop Radiosonde Worksheet

Some universities will have the facility to launch a weather balloon (Fig. 5.10). The procedures for doing so will vary depending on the equipment available and are not provided here. For an overview of such procedures please go to http://www.weathergraphics.com/tim/raob/. This worksheet is provided to record relevant information during a balloon flight.

1. Date:

2. Time of Launch:

3. Weather conditions at time of launch (temperature, dewpoint, wind speed and direction, sky cover and cloud type(s):

4. Estimated ascent rate (from data on terminal readout):

5. Minutes elapsed before balloon entered cloud base (if present):

6. Estimated cloud base height, based on elapsed time:

7. Pressure, temperature, and dewpoint at cloud base (if present):

8. Peak wind observed during sounding (direction, speed, pressure level, and estimated altitude):

9. Time and final pressure at end of sounding (if known):

10. Circumstances of termination (if known -- e.g., balloon burst, lost signal, etc.):

Chapter 6 Wind, the General Circulation, and Storms

'Big whorls have little whorls, which feed on their velocity.
And little whorls have lesser whorls –
and so on to viscosity.' – Lewis F. Richardson

The term wind is generally used to describe the horizontal motion of air. Vertical air motions usually are called updrafts or downdrafts. Wind velocity has both a direction and a magnitude, therefore, it is a *vector* quantity. Wind speed is measured with an instrument called an *anemometer*. Most often, winds in the lower atmosphere have speeds of a few meters per second (m/s), but in the upper atmosphere they often exceed 50 m/s (112 mph), and in tornadoes and in hurricanes they may exceed 100 m/s (224 mph). Wind direction is measured with a *wind vane* and is given as the direction *from* which the wind is blowing.

In the eighteenth century, Sir Isaac Newton discovered three laws of motion:

I)   In the absence of forces, an object at rest will remain at rest and an object in motion will remain in motion with the same velocity.

II) The acceleration of a body is equal to the force acting on it divided by its mass.

III) To every action there's an equal and opposite reaction.

Based on these laws, the velocity of the air usually can be calculated if one knows the forces acting on it.

There are five forces governing winds.

1) *pressure-gradient force* - occurs because atmospheric pressure varies from place to place.

2) *gravity* - only acts in the vertical direction and is closely balanced by the vertical gradient in pressure. This balance of forces is called hydrostatic balance.

3) *Coriolis force* - exists because the Earth rotates under a moving air stream, resulting in an apparent deflection of the wind as observed from the ground. In fact, the Coriolis force is not a true force in the sense of the pressure-gradient force or gravity. It is sometimes called the Coriolis effect. If the wind were measured with respect to a point fixed in space there would be no Coriolis force. In such a circumstance, an air parcel with no forces acting on it would move in a straight line at a constant speed. But we measure wind velocity with respect to the surface of the Earth. From this reference frame, the effect of the Earth's rotation is equivalent to a force. The following observations summarize the effects of the Coriolis force: a) It is proportional to the speed of the moving object, b) is zero at the equator and increases to a maximum at the poles, and c) always acts at 90° to the motion, resulting in a deflection of air parcels to the right of their motion in the Northern Hemisphere and to the left in the Southern Hemisphere. In the absence of other force, air parcels would display a curved path to the right in the Northern Hemisphere.

4) *centrifugal force* - is an apparent force that is invoked when the wind follows a curved path such as those generally found in cyclonic storms. Newton's first law of

motion states that a parcel in motion will continue in a straight line unless acted upon by an unbalanced force. In the vicinity of a low-pressure center in the atmosphere, the orientation of the pressure gradient force changes as the flow moves around the low. This provides the unbalanced force that causes the wind to follow a curved path. The centrifugal force is a fictitious force that acts opposite this changing pressure gradient force and actually reflects the air's inertia or tendency to move in a straight line.

5) *frictional force* - is present in all moving systems and acts to oppose the motion. In the atmosphere, friction is most important near the surface (lowest ~1 kilometer of the atmosphere) and causes dissipation of the energy of motion into heat energy.

When isobars are straight and frictional forces are negligibly small, the wind velocity is determined only by the pressure-gradient and Coriolis forces. In such cases, the pressure gradient and Coriolis forces balance each other, and the resultant wind is said to be geostrophic. The geostrophic wind is parallel to the isobars with low pressure on the left if you stand with your back to the wind in the Northern Hemisphere. The closer the isobars are to one another, the higher the wind speed. In the Southern Hemisphere, the same result occurs, except that low pressure is on the right when the wind is at your back.

When isobars are curved, such as they are around a circular region of low pressure, the effects of curvature must be taken into account. When the air is moving along a curved path, the net force must be producing an acceleration towards the center of rotation. When the wind is blowing parallel to curved isobars under the action of the pressure gradient, centrifugal and Coriolis forces, it is called the *gradient wind*. The wind is nearly in gradient-wind balance at altitudes above one kilometer, where frictional effects of the ground are small.

Frictional forces act in a direction opposite to the direction of the wind-velocity vector. They act to reduce the velocity, which, in turn, reduces the Coriolis force. As a result, the wind is deviated across the isobars towards low pressure. The frictional effects are a maximum near the ground and decrease with height. In the Northern

Hemisphere, when the directions of the isobars do not change with height, the wind vector turns clockwise with height up to about one kilometer, where the winds are in gradient or geostrophic balance. The rotation of the wind vector near the ground sometimes is called the *Ekman spiral.*

Along coastlines, wind velocities often deviate markedly from the geostrophic or gradient values. Differential heating of land and water can produce air temperature and pressure differences, and a convection cell is established. Cool air from over the water moves over the land in the form of a *sea breeze.* Aloft, air moves over the sea and sinks, providing a good example of a convection cell. Similarly, at night the land cools faster than the water surface, and a *land breeze* is established. In a sense, monsoons are sea and land breezes on a massive scale. In the winter, cool, dry continental air tends to flow towards warmer oceans; in the summer, warm, humid oceanic air blows over the land.

In hilly and mountainous regions, special winds arise as air sinks or rises along the slopes. The *Foehn, Chinook, and Santa Ana* winds flow downhill in response to the larger-scale circulation and are warm and dry. In desert areas where radiation effects cause pronounced warming and cooling, mountain and valley winds are common. During the day, winds blow up the sun-heated valleys; at night, cool, heavy air sinks down the valleys. Special winds also result when cold, dense air originating over elevated (often over ice) plateaus descends to lower elevations. Such winds are called *katabatic* winds and are pervasive along the coasts of Greenland and Antarctica.

General wind circulation of the Earth's atmosphere

The atmosphere can be regarded as a heat engine, which receives energy from the sun and converts it into the kinetic energy of air motions. Of the vast energy received on the average from the sun, only two to three percent of it can account for the kinetic energy represented by the wind systems over the Earth.

As noted in Chapter 3, at high latitudes the Earth radiates more energy to space than it absorbs, while at low latitudes it absorbs more than it emits. As a result, there must be heat transported poleward in order to prevent long-period temperature changes. Ocean and air currents accomplish most of the heat transport. If the Earth were a uniform stationary object, you would expect a huge convection cell with air rising at low latitudes, moving poleward aloft, sinking at high latitudes, and returning equatorward near the ground. You would also expect a convection cell between the day side and night side of the Earth. But the Earth is a non-uniform, rotating spheroid, and as a result the air motions are not in the form of simple convection cells.

Interestingly, there is a convection cell, called the *Hadley cell*, at low latitudes. Air rises over equatorial regions, moves poleward aloft, sinks at latitudes of about 30°, and flows equatorward near the ground in the form of the trade winds (see Fig. 6.1). In the Northern Hemisphere, the trades blow mostly from the northeast; in the Southern Hemisphere, from the southeast. The trade winds converge along the *intertropical convergence zone* (ITCZ) or equatorial trough, where winds are light and the air is warm, humid, and uncomfortable. This region is sometimes called the *doldrums.*

In the latitude belt near 30°, where sinking air predominates, large, semi-permanent anticyclones prevail. Within them, the air tends to be dry and the winds light over very large areas. These areas were named the *horse latitudes* by the sailors of long ago.

Poleward of the Hadley cell, there is a broad latitude region dominated by westerly winds that increase with height and have embedded within them traveling cyclones and anticyclones. On the average, in the belt of westerlies, the winds near the ground have a component away from the equator and a reverse component aloft. Poleward of about the 60° latitude circle, there is a third cell with easterly winds predominating near the surface.

Observations of pressure and wind patterns reveal significant variations from season to season. The wind fields follow the sun. For example, the ITCZ has an average position at about 5° S latitude in January and moves northward as the months go by, reaching a position averaging about 10°N latitude in July.

The strength of the winds, as noted earlier, depends mostly on the pressure-gradient force. In the winter, the pressure-gradient force between latitudes of the westerly winds (about 30° to 60°) is at a maximum. This occurs because the north-south temperature gradient is a maximum. As a result, the westerly winds are strongest during that season. The highest velocities tend to be concentrated in a restricted current called the *polar-front jet stream,* found at altitudes of about 12 km, over the portion of the polar front across which north-south temperature changes are concentrated. The polar-front jet stream exhibits average maximum winds of about 60 m/sec, but can be much stronger. It tends to follow a meandering path around the globe at middle latitudes. A second strong westerly wind current, the *subtropical jet stream* is found at lower latitudes and slightly higher altitude than the polar-front jet.

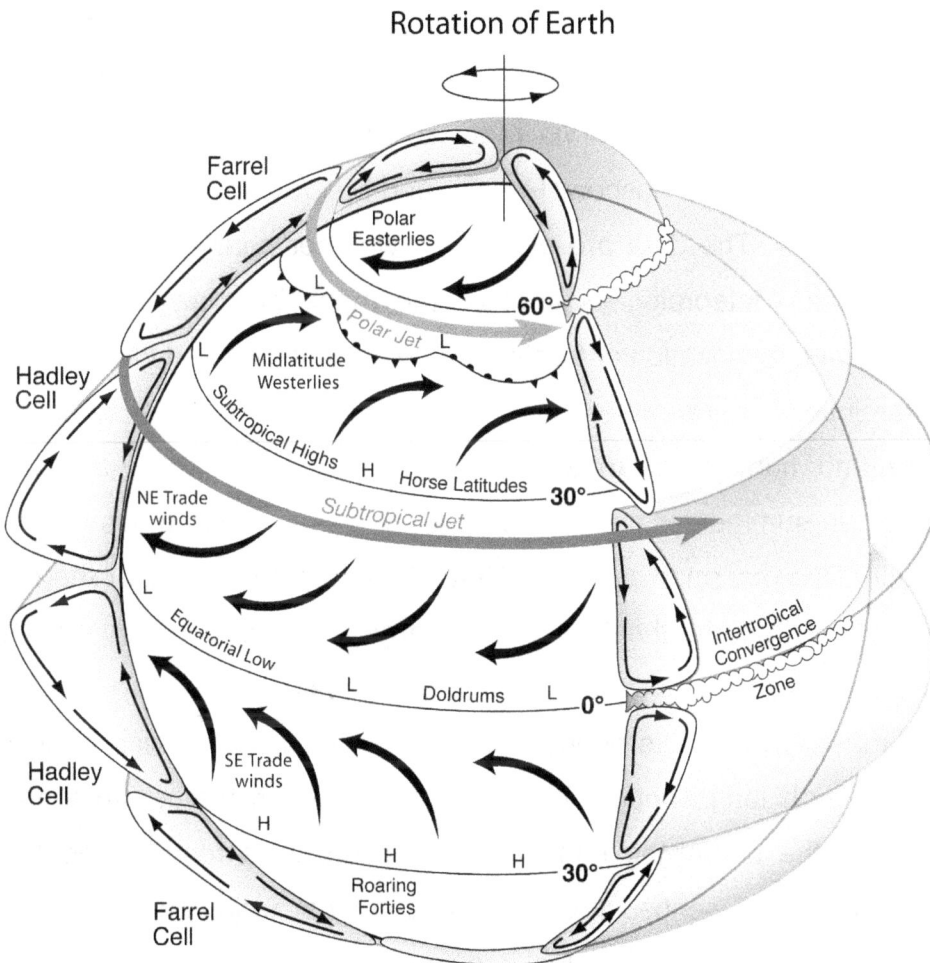

Figure 6.1 General circulation of the atmosphere

Jet streams, particularly the one over the polar front, have important effects on the global weather patterns by contributing to the transport of atmospheric properties and the formation of storm centers. Jet streams have both good and bad influences on aviation. By planning flights to maximize tail winds, airplane ground speeds can be optimized in order to reduce flight times. On the other hand, in regions where the wind-velocity gradient is large, that is, where the *wind shear* is large, there can be strong *clear-air turbulence* (CAT). Pilots seek to increase ground speed while avoiding turbulence.

Air masses, fronts and midlatitude cyclones

The state of the atmosphere can be depicted on synoptic maps, that is, maps showing patterns of various weather elements at a particular time. Surface weather stations and rawinsonde stations all over the world follow observing practices formulated by the World Meteorological Organization (WMO) in Geneva, Switzerland. The observations are transmitted via radio and computer to central receiving points such as Washington, D.C., Moscow, Russia, and Melbourne, Australia, and then on to many other stations. The data are plotted on weather charts according to standard WMO procedures. Meteorologists analyze the maps by drawing such features as isobars, isotherms, the location of fronts, and centers of high and low pressure.

An examination of a weather map shows that there are large regions over which temperature and humidity change relatively little. For example, the southeastern United States, in summer, is often covered by a warm, humid mass of air moving in from the Gulf of Mexico and Caribbean Sea. At the same time, over the northeastern United States and Canada, there could be a large region of cool, dry air. The term *air mass* is used to represent these distinctive bodies of air.

An air mass develops its principal characteristics by remaining over a particular region of the Earth long enough for the air to approach temperature and moisture equilibrium with the underlying surface. This occurs as a result of heat transfer by various mechanisms (described in Chapter 2) and by evaporation or condensation.

Air masses are classified mostly according to the region of formation and whether or not the underlying surface is oceanic (maritime) or continental. Four principal air masses and the letters used to designate them are:

i) maritime tropical—mT   iii) maritime polar—mP

ii) continental tropical—cT   iv) continental polar—cP

Two additional air mass categories sometimes referred to are *arctic* and *equatorial* corresponding to air masses that form over polar ice sheets at high latitude, and over equatorial rain forests, respectively. Once an air mass leaves its source region, it can be modified by exchanges of heat and moisture with the surface over which it passes. For example when a Canadian continental-polar air mass passes over the warm waters of the Gulf Stream it is rapidly modified by the transfer of heat and moisture to the air from below.

It is found that when air masses having significantly different properties encounter one another, they do not mix readily. Instead, the cooler air wedges itself under the warmer air, and a zone of transition between them that is called a front separates the two air masses.

There are various types of fronts, whose names depend mostly on the direction of movement:

Cold front—the cold air advances, replacing warm air.

Warm front—the warm air advances, replacing cold air.

Stationary front—the transition zone between cold, and warm air remains essentially stationary.

Occluded front—the frontal system resulting when a cold front overtakes a warm front, forcing the warm air aloft.

Frontal zones are responsible for the formation of much of the cloudiness, rain, and snow that occur over the United States, especially in winter. The warm, humid, less dense air is forced to rise over the cold air. In the process, condensation and precipitation take place.

Frontal zones also are important because they are favored locations for the formation of *cyclones.* A cyclone is a region of low pressure around which there is closed circulation. In the central United States, the term cyclone is sometimes used to mean a tornado; in southeast Asia, the name cyclone refers to tropical storms in the hurricane family. When a meteorologist speaks of midlatitude cyclones, he is talking about regions of low pressure several hundred to a thousand kilometers in diameter that form in middle latitudes. Midlatitude cyclones, more commonly known as winter storms, form along pre-existing fronts and are the source of a great deal of clouds, rain, and snow.

Midlatitude cyclones develop along frontal zones because denser, cold air is located at the same height as nearby, less-dense, warm air. Such a condition represents a source of potential energy. The heavy air sinks, displacing the warm air, and converts potential energy into kinetic energy in the form of a cyclonic wind circulation.

The following life cycle is often seen on weather maps (Fig. 6.2). A small wave develops on a stationary front. At the surface, pressures begin to fall. A continuation of this process for a day or two leads to the formation of a mature cyclone whose fronts have an open wave character. As the wave grows, it takes on the appearance of a breaking wave with a cold front advancing and overtaking a warm front. The cyclone may last for more than a week, growing in size and intensity as the central pressures fall. Such storms may become > 1000 km in diameter and produce heavy rain, snow, and strong winds. Finally when the warm air is carried aloft and the cold air has descended near the surface that storm dissipates.

Midlatitude cyclones occur most frequently in regions where major frontal systems tend to stagnate. For example, in the winter, polar air from North America pushes far to the south. The associated front tends to stagnate along the eastern slope of the Rockies and through the Gulf of Mexico. Cyclone initiation is common over Colorado, Alberta, Canada, and over the Gulf of Mexico. Off the east coast of the United States the proximity of the warm Gulf Stream to the colder land creates especially favorable conditions for cyclogenesis.

The polar front extends across the Pacific and Atlantic Oceans and defines a preferred belt of cyclonic development. Many cyclones that eventually affect North America originate in the north Pacific, in the vicinity of the Aleutian Islands. They account for a semi-permanent center of low pressure called the Aleutian low. It has an Atlantic counterpart known as the Icelandic low.

A special group of cyclones are called thermal lows. They occur in summer over hot regions such as the deserts of Arizona and northwestern Mexico. As the air is warmed air molecules take up more space with the result that the surface pressure is decreased. Thermal lows are very shallow and their circulation weakens rapidly with

144

height. Another class of cyclones, those that form in the tropics will be discussed in Chapter 8.

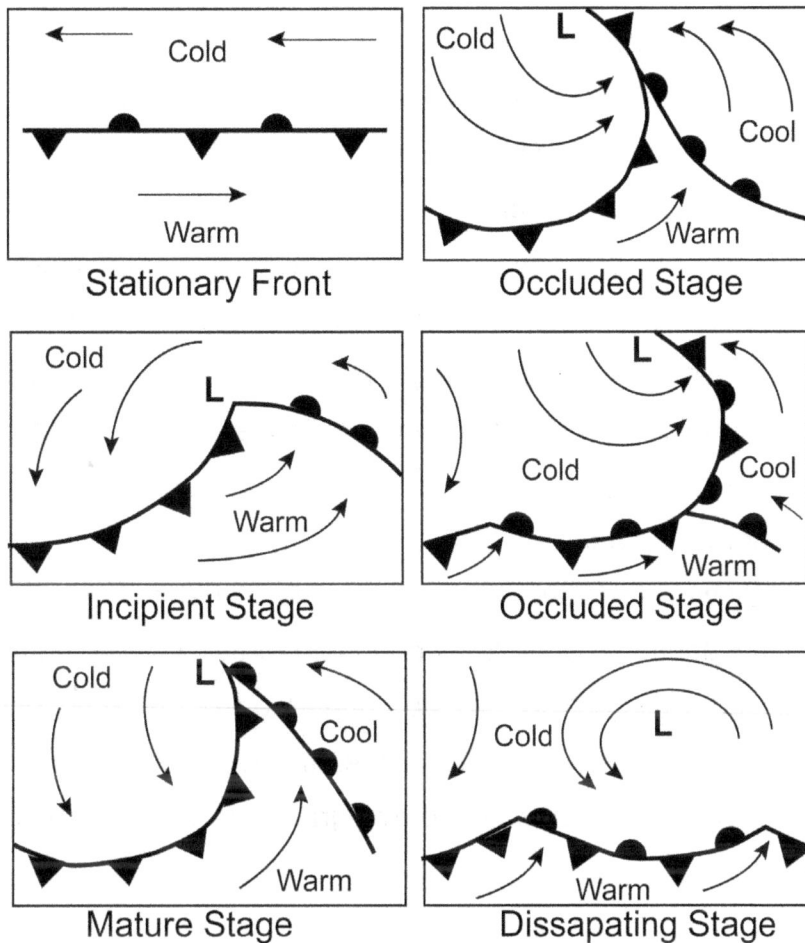

Figure 6.2 Schematic of the evolution of a midlatitude cyclone or winter storm.

Lab 26: Wind: Air in Motion

INTRODUCTION

Winds arise from pressure gradients. As areas of high and low pressure migrate, evolving pressure gradients between them will force the air in regions of higher pressure to move toward regions of lower pressure, creating winds. Once the air is moving, other forces come into play. The Coriolis force, due to the rotation of the Earth under moving air, prevents the air from taking a path directly toward lower pressure, thus contributing to the spiral wind patterns seen in storm systems.

When a balloon is blown up, the pressure of the air in the balloon becomes greater than the pressure of the air surrounding the balloon. Therefore, when the balloon is opened, the high-pressure air rushes out into the surrounding region of lower pressure producing a puff of wind.

There are, however, differences between the model of wind in this activity and winds in the atmosphere. The material boundary of the balloon that gives rise to the pressure gradients in this activity has no counterpart in the atmosphere. Thus, the creation of wind is usually not as abrupt in the atmosphere as when the balloon being released. The acceleration of wind occurs gradually as a storm develops and its central pressure drops.

ACTIVITY

OBJECTIVE: The objective of this activity is to investigate the processes involved in creating wind.

MATERIALS:
§ balloon (Long ones work best, as opposed to round ones)
§ string or fishing line (5 meters)
§ drinking straw
§ clear tape

PROCEDURE:

1. Thread the string through the straw and have one person hold one end and you hold the other.

2. Blow the balloon up, but do not tie it. With the help of your partner, tape the balloon to the straw (Fig. 6.3).

3. Pull the string tight and move the straw to one end of the string.

4. Let go of the balloon and observe what happens. Record your observations.

5. Repeat the process two more times. Record any additional observations.

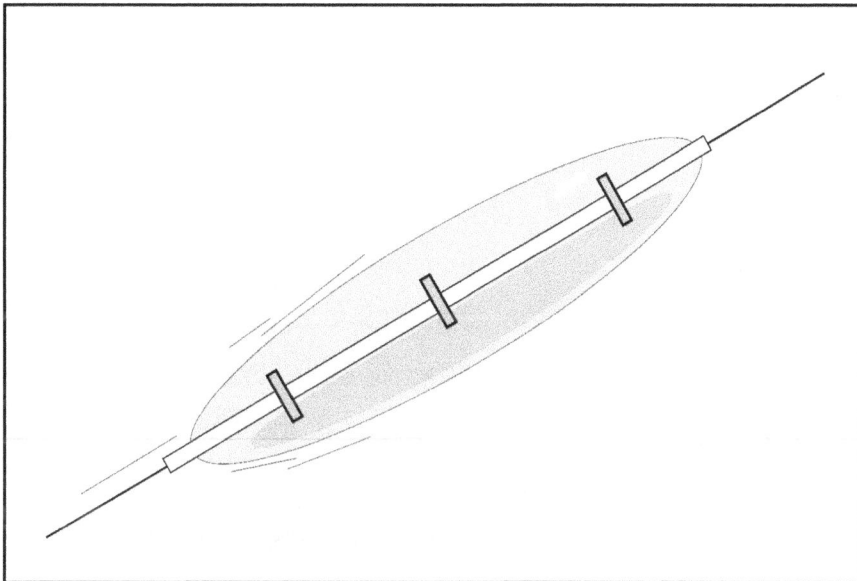

Figure 6.3 Schematic diagram

QUESTIONS:

1. What happened to the air in the balloon when the balloon was released? Why?

2. In your trials, what force(s) caused your balloon to slow or stop?

3. What forces that are generally important for winds in the atmosphere are not important here?

Lab 27: The Coriolis Effect

INTRODUCTION

As winds and ocean currents move across the Earth's surface, they cover long distances over long periods of time and experience a deflection due to the rotation of the Earth underneath the air. The farther from the equator the motions occur, the greater the deflection is. The deflection causes winds flow in a counterclockwise direction around low-pressure centers in the Northern Hemisphere and in a clockwise direction in the Southern Hemisphere.

Imagine that you and a friend are standing inside a large box that is mounted on a carousel. If the carousel is slowly rotating, you can imagine not being aware of it, just as you are unaware that the Earth is rotating daily on its axis. If you and your friend begin throwing a softball back and forth, it will appear to be deflected to one side, as if pushed off course by some unseen force, making it hard to catch.

A similar unseen force acts on moving air currents in the atmosphere. It is called the "Coriolis" force and the deflection it causes is the Coriolis effect. Like centrifugal force, the Coriolis force is only an apparent force and the Coriolis effect is only an apparent deflection. Yet both are important in understanding the motion of air parcels in the atmosphere.

Although the Earth is rotating, we do not take the effects of this rotation into account in our daily lives because it is relatively slow. For example, when playing baseball we do not notice a deflection of the ball due to the rotation of the Earth even though it does occur. That is because a baseball travels across the field quickly, not allowing the Earth to rotate very far underneath the ball during this short period. Since winds and ocean currents travel over long periods and distances the effect becomes important and can be seen in the structure of atmospheric storm systems and in the ocean circulation.

ACTIVITY

OBJECTIVE: The purpose of this activity is to investigate how objects are deflected in a rotating frame of reference.

MATERIALS:

§ Obtain a globe preferably with a diameter of 14" or larger that can freely rotate on its axis (Fig. 7.4).

§ Cover the Northern Hemisphere of globe with cellophane. Use short overlapping pieces to keep the surface smooth.

§ Overhead marking pen (washable, not permanent ink)

§ Moist sponge or towel to clean pen marks from cellophane

PROCEDURE:

1. With the globe stationary, start near the North Pole, and lightly draw a line moving your hand straight south toward the equator.

2. With the globe stationary, start near the equator, and lightly draw a line moving your hand straight north toward the North Pole.

Figure 6.4 Schematic diagram of globe

3. Predict the shape of the lines when the globe is rotating while the lines in steps 5-8 are being drawn.

4. Spin the globe to obtain a steady moderate rotation (clockwise looking down on the North Pole).

5. Starting near the North Pole, lightly draw a line moving your hand straight south toward the equator.

6. Starting near the equator, lightly draw a line moving your hand straight north toward the North Pole.

7. Starting at the Equator make a straight horizontal motion with the pen from left to right.

8. Starting at 45° N latitude make a straight horizontal motion with the pen from left to right.

9. Repeat steps 3-8 with a counterclockwise rotation of the globe.

QUESTIONS:

1. Describe the path of the ink lines in the case of no rotation?

2. How did you do in your prediction of the line shapes given a rotating globe?

3. How does the direction of travel of the pen, north vs. south affect the resulting line shape?

4. Compare the shape of the lines in Steps 7 and 8. What is the difference?

5. How does a change in the direction of rotation affect the resulting shape of the lines?

6. When you turn the demonstrator in a counterclockwise direction, which hemisphere is it representing? And if you turn it clockwise, which hemisphere is represented?

7. Can you formulate your observations into a set of general rules describing the impact of rotation on the path of air parcels in the free atmosphere?

ALTERNATIVE CORIOLIS DEMONSTRATOR

§   Obtain or construct a Coriolis demonstrator that has a freely rotating base. A "Lazy Susan" or a round swivel stool will work.
§   Attach a plate or "magic slate" to the rotating base. The plate ideally should be about 18-24" in diameter and have a surface that can be marked (Fig. 6.4).
§   A ball bearing (inch) or a marble
§   Chalk or other writing implement to mark the surface of the plate

PROCEDURE:
Roll a ball bearing or marble across the surface of the plate with no rotation. Spin the top of the rotating plate slowly. Draw a chalk line directly across the rotating plate. Even though you moved the chalk in a straight line, the mark was curved. Experiment with spinning the plate in different directions. Or start your mark from the middle of the rotating plate.

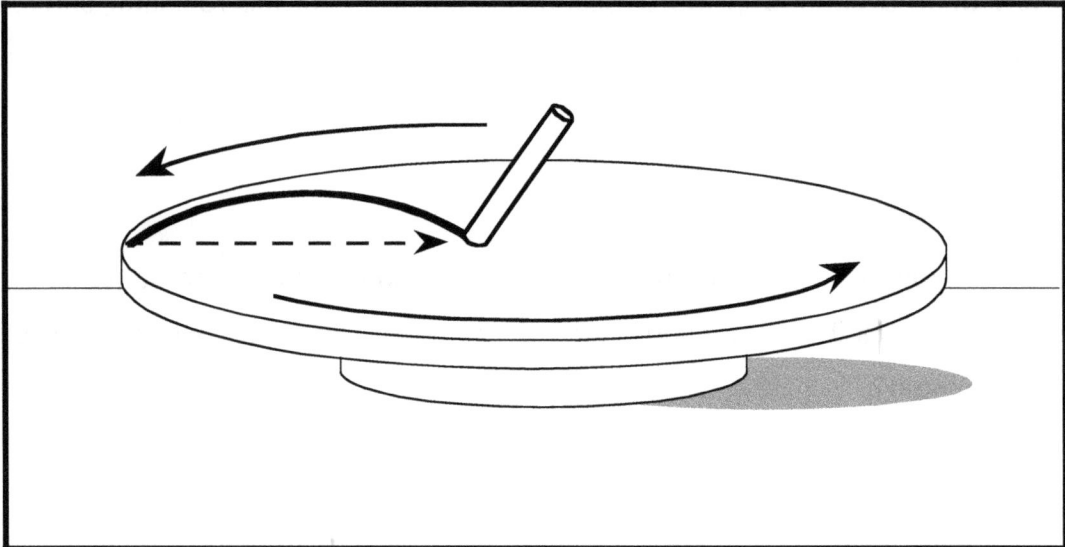

Figure 6.5 Schematic diagram

QUESTIONS:

1. Describe the path of the ball bearing in the case of no rotation?

2. Rotate or spin the plate in a counterclockwise direction and roll the ball bearing across it. Describe the ball bearing's motion.

3. Predict the motion of the ball bearing if the plate was rotated in a clockwise direction.

4. Rotate the demonstrator in a clockwise direction. How does your prediction compare with what happened?

5. The center of the plate represents the poles of the Earth. Predict what would happen if you released the ball bearing from the center of the plate when it was rotating clockwise. Predict what would happen if you released the ball bearing from the center of the board when it was rotating counterclockwise.

6. What happened?

7. Release the ball bearing from the center of the plate when it is rotating counterclockwise. What happened?

8. When you turn the demonstrator in a counterclockwise direction, which hemisphere is it representing? And if you turn it clockwise, which hemisphere is represented?

Lab 28: Forces, Wind, and Weather Maps

INTRODUCTION

Specific forces control the movement of air in the atmosphere. The exercises in this lab explore the relationship between these forces and the resulting winds as represented on weather maps.

ACTIVITY

OBJECTIVE: Examine how combinations of atmospheric forces balance, leading to the air motions we observe.

QUESTIONS:

1. Suppose the air at **X** is initially at rest and then a pressure field given below, is introduced causing the air to move. Indicate with an arrow the direction that the air will move if no other forces are acting on it.

        L   **x**   **H**

2. In Fig. 6.6 below, height contours at 700 mb (above the friction layer) show regions of high and low pressure in the Northern Hemisphere. Sketch a series of arrows to indicate the direction the wind blows around the areas of low and high pressure.

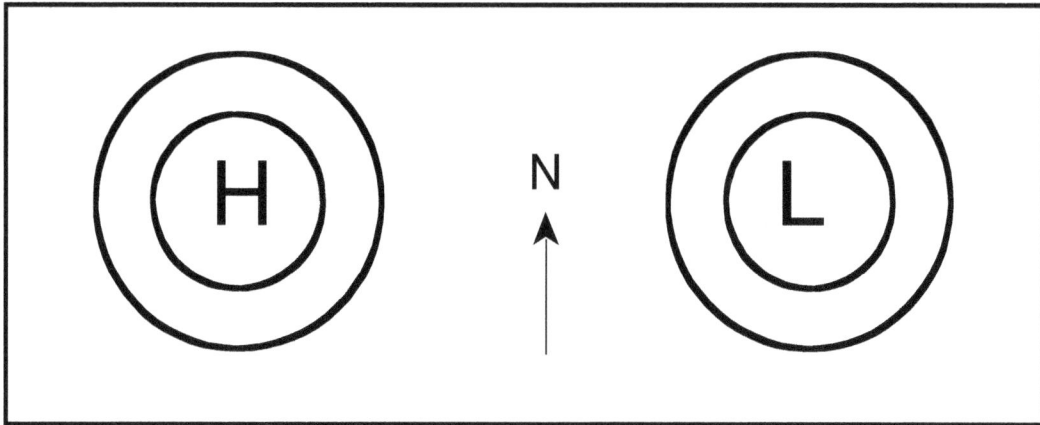

Figure 6.6 Schematic diagram

3. a. On Fig. 6.7 below sketch wind vectors for each of the labeled points. The wind vectors are arrows that point in the direction of the wind and whose length indicates the relative wind strength.

b. Are the winds stronger at point A or point C? Explain.

c. Sketch a solid line indicating the ridge axis, or axis of higher heights.

d. Sketch a bold dashed line showing the trough axis, or axis of lower heights.

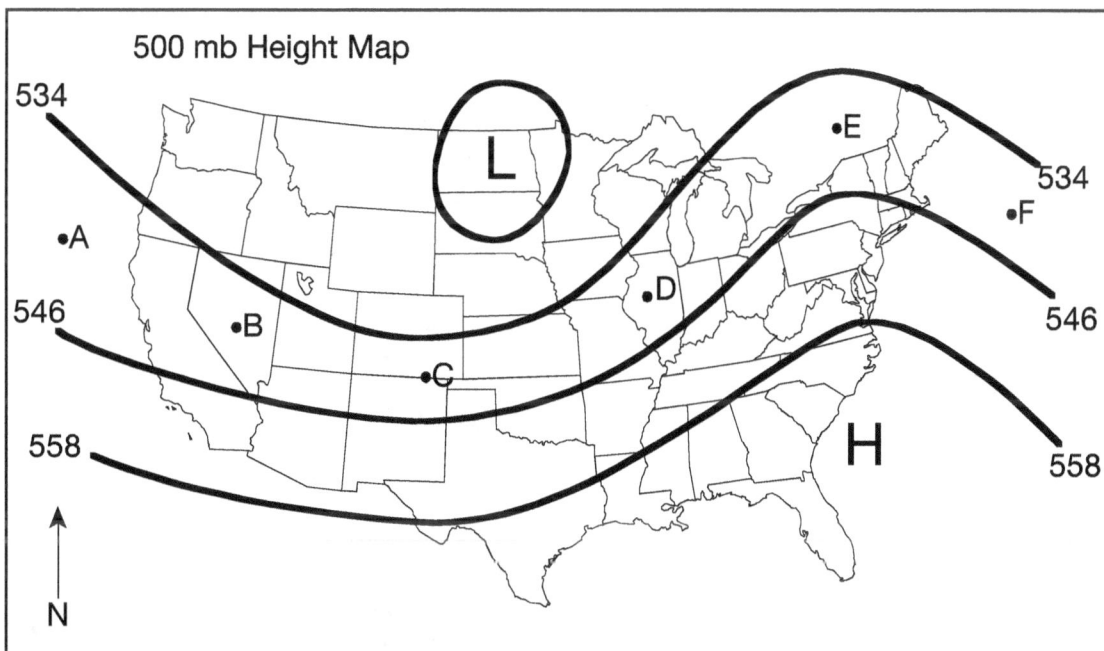

Figure 6.7 Map showing contours of constant height (tens of meters) above sea level of the 500-mb pressure surface.

4. Indicate the forces and wind direction at point **X** below assuming:

   a) A gradient wind balance

   b) A frictionally modified gradient wind model

5. Each of the panels in Fig. 6.8 below represents the wind pattern around a region of low or high pressure.

   a. Below each figure indicate which hemisphere (NH or SH) the figure is in.

   b. Indicate whether the circulation seen is a center of low or high pressure.

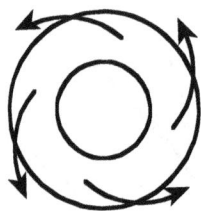

i. _____   ii _____

iii. _____   iv _____

Figure 6.8 Schematic diagram

6. a. Friction induces horizontal motion into a surface low and out of a surface high pressure center (e.g., Fig. 6.8). Air cannot accumulate indefinitely in a low (it would eventually become a high if it did), or continuously flow out a high (it would eventually become a low). Therefore, what does the air flow have to do above a surface low to keep it form filling up with air? What does it have to do above a surface high?

b. Indicate whether the air above each panel in Fig. 6.8 above will be rising or sinking.

7. On Fig. 6.9 below sketch wind vectors for each of the four labeled points. The wind vectors are arrows that point in the direction of the wind and whose length indicates the relative wind strength.

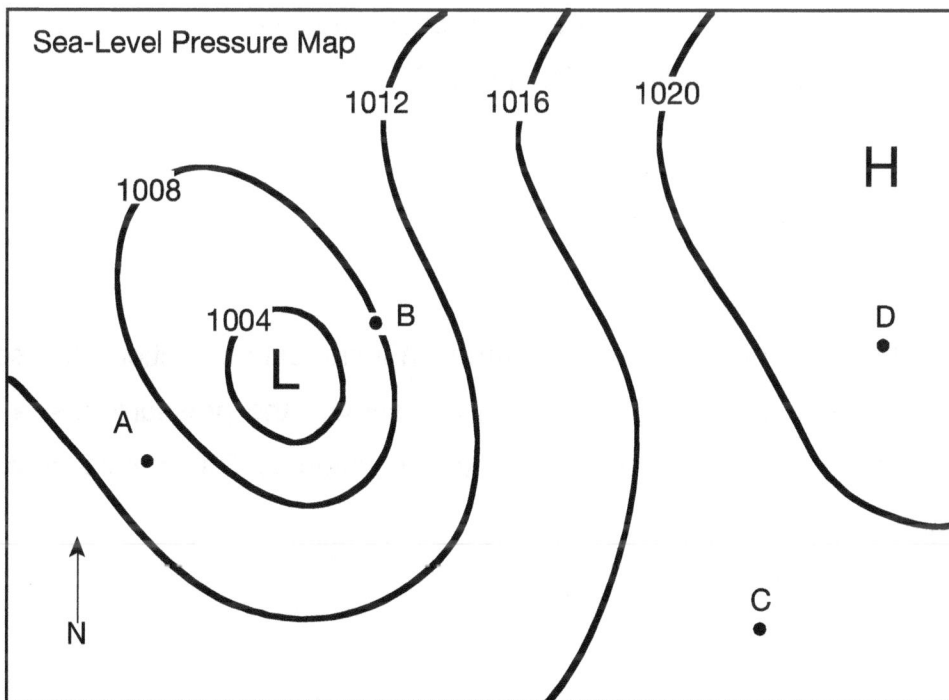

Figure 6.9 Sea-level pressure map showing isobars (mb).

a. What is the sea-level pressure at point B?

b. Toward which direction is the wind at point A blowing?

c. The pressure gradient force at point B is directed toward which point? L, C, H, D?

d. Toward which direction is the wind at point C blowing?

e. Toward which direction is the wind at point B blowing?

f. At which point is the wind strength the greatest?

g. At which point is the wind strength the weakest?

8a. In a standard atmosphere, the air exerts a force of 14.7 pounds per square inch downward on the ground at sea level. This pressure is due to the gravitational force of the earth pulling the atmosphere downward. What force, directed upward, balances this gravitational force?

b. The change in atmospheric pressure with increasing elevation near sea level is approximately 10 mb per 100 meters. Estimate the pressure at sea level give the surface pressure is 955 mb at an elevation of 500 meters (see Fig. 6.10 below).

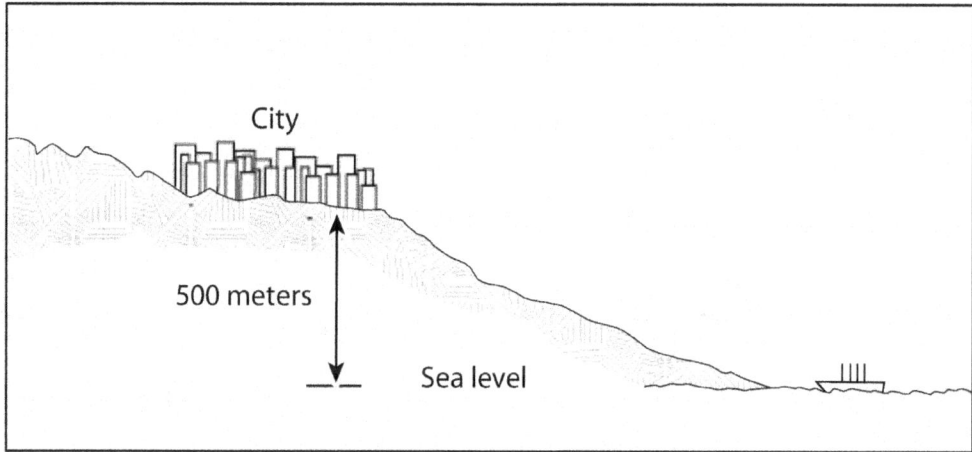

Figure 6.10 Schematic diagram

Lab 29: Jet Stream Winds

INTRODUCTION

The term *jet stream* is used to refer to the rivers of wind high in the atmosphere, above about 20,000 feet, that steer storms. These winds not only steer storms, but are also intimately involved in the development of areas of high and low air pressure at the Earth's surface. The American Meteorological Society's *Glossary of Weather and Climate* defines a jet stream is "a relatively narrow river of very strong horizontal winds (usually 50 knots or greater) embedded in the planetary winds aloft."

Jet streams form along the upper-air boundaries of large masses of warm and cold air. Warm air molecules take up more space than cold air molecules, because of their higher kinetic energy. Therefore, when warm and cold air masses collide, as in a winter storm, the surfaces of constant pressure slope increasingly with altitude, resulting in an pressure gradient force and winds that strengthen with height (Fig. 6.11). The largest pressure gradients (steepest isobar slopes) are located at or near the tropopause, thus the strongest winds are found there. The slope reverses in stratosphere where warmer air resides over the poles.

During major cold outbreaks over the USA, the jet stream dives south, staying above the warm-cold boundary, and sometimes moving well over the Gulf of Mexico. The jet stream retreats northward into Canada during unusually mild winter weather and during the summer.

One of the things that meteorologists look for are the presence of *jet streaks*, smaller cores of higher winds embedded in jet streams (Fig. 6.12). These are particularly important weather makers, linked to the formation of blizzards and tornadic thunderstorms alike because of the pattern of rising and sinking air associated with them.

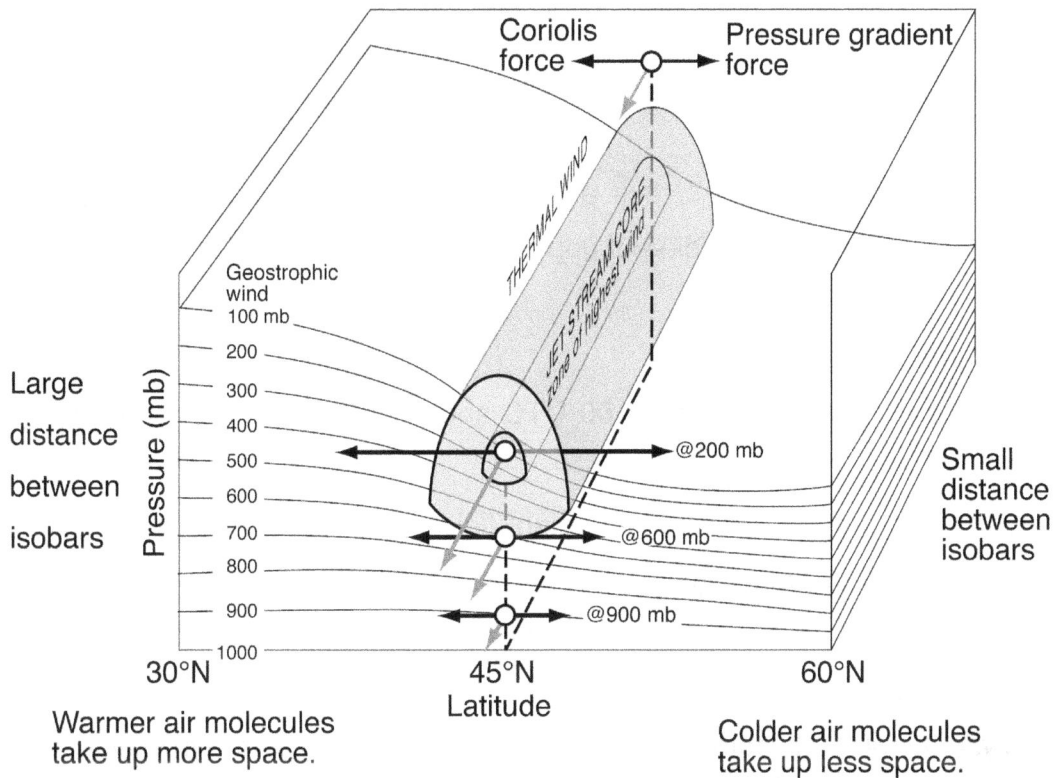

Figure 6.11 Schematic diagram of *thermal wind balance* in which the pressure gradient force equals the Coriolis force under conditions of hydrostatic balance.

The structure of the jet stream can be studied in upper-level charts. On upper-level charts, it is the height (in meters) of a given pressure above the surface that is plotted rather than the pressure. Therefore, an upper-level chart is literally a map of elevation contours showing the height in the free atmosphere of the given pressure surface (which also give its name to the chart; e.g., 500 mb chart).

Figure 6.12 Schematic diagram of a jet streak and pattern of rising and sinking air associated with the entrance and exit regions of the jet streak. The shading indicates regions of increasing wind speed.

ACTIVITY

OBJECTIVE: The purpose of this activity is to investigate the structure of the jet stream, through an understanding of upper level charts.

PROCEDURE:

Attached are the 500 mb charts at 00 UTC and 12 UTC on 15 December 1987 (Fig. 6.13).

1. Analyze the 500 mb geopotential height field with a contour interval of 60 m using standard contour values (divisible by 60; e.g., 5340, 5400, 5460,.....5640, 5700, 5760, etc.).   Finalize these contours in dark pencil and label them (with the last zero dropped; e.g., 534, 540, etc).

2. Analyze the 500 mb isotachs with a contour interval of 20 kt. Only include isotachs for wind speeds of 30 kt and higher (30, 50, 70,...). Finalize these contours in blue pencil. (Note that wind barbs = 10 kt and flags = 50 kt on the charts)

3. Shade regions of the chart with wind speeds > 70 kt in light blue.

4. Mark the estimated position of the maximum wind speed on each chart with a red "X". Draw a red line on each map to indicate the axis of maximum winds in the jet streak in the area with wind speeds > 70 kt.

5. Estimate the speed of propagation of the jet streak (kt) from the distance that the red "X" moves in 12 hours. To estimate the propagation speed use the approximate conversion factor that 10° latitude = 600 nautical miles[2] =~45 mm on the charts.

QUESTIONS:

1. What is the relationship of the wind direction and the contours of constant height in your analyses? Comment?

---

[2] Note that 1 kt = 1 nautical mile per hour. Sixty nautical miles = 1 degree latitude.

2. What is the relationship between the wind speed and the spacing between the contours of constant height in your analyses? Comment?

3. Is the propagation speed of the jet streak greater or less than the maximum wind speeds in the jet? Comment?

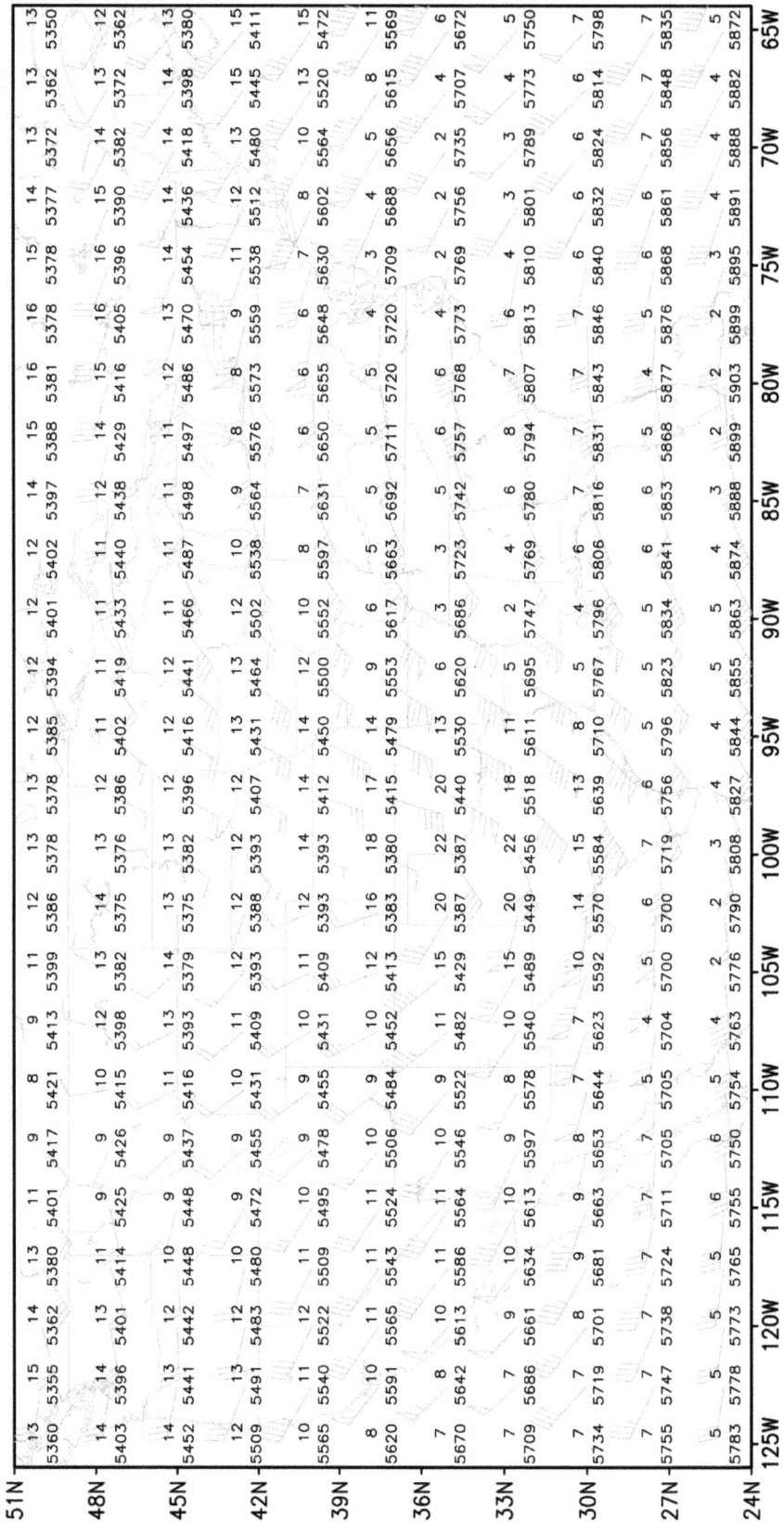

Figure 6.13a 500 mb chart for 0000 UTC on 15 December 1987.

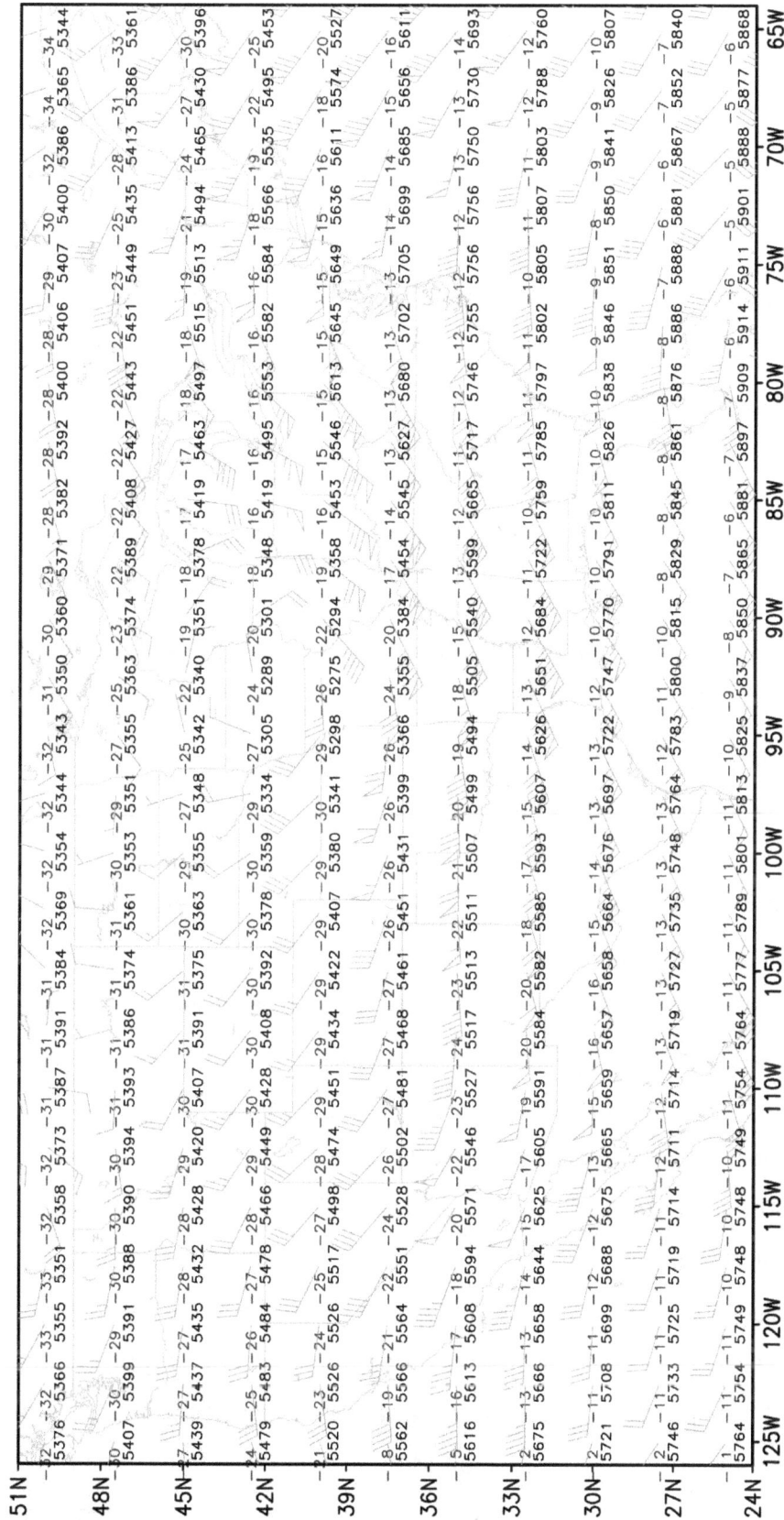

Figure 6.13b 500 mb chart for 1200 UTC on 15 December 1987.

Lab 30: Analyzing a Winter Storm

INTRODUCTION

The lab explores the surface structure of an explosive cyclone that developed over the Mid-West United States during 14-15 December 1987. Research shows that explosive cyclogenesis (a drop of central pressure of more than 24 mb in 24 hours) rarely occurs over land. However, this storm's development occurred entirely over land. During an 18 hour period of deepening the cyclone experienced a central pressure drop of 27 mb, and within that time period the cyclone experienced a six-hour pressure drop of 15 mb.

A large portion of the nation experienced blizzard conditions, heavy snow, record temperatures, high winds, and severe weather during the course of the storm's rampage . O'Hare International Airport, experienced its all time record low surface pressure (981 mb) when the storm was still 100 km to the southwest. The storm system produced an extensive band of heavy snow extending from the southern Rockies northeastward towards the Great Lakes. Snowfall totals of a foot or more were common throughout this area. El Paso, Texas set a new 24-hour snowfall record of 16.8 inches fell. Kansas City, Missouri broke their 24-hour snowfall records for December by receiving a foot of new snow.

Record high temperatures for the month of December were experienced in the southeast with a record high of 81°F observed in Baton Rouge, Louisiana. To the north and west, exceptionally cold weather resulted in record low temperatures as cold air poured out of Canada in the wake of the advancing cyclone. Lubbock, Texas broke their previous record low of 10°F, set in 1917, when the mercury fell to 1°F.

Damaging high winds were felt from the Rockies eastward to the Appalachians causing extensive damage. In Illinois, wind gusts of approximately 35 m s^{-1} were observed. High winds downed trees and power lines causing power outages effecting over 165,000 people in the Chicago area alone. For just the fourth time in the past 20 years O'Hare International Airport halted operations due to inclement weather.

Additionally, severe thunderstorms were spawned as the cyclone matured. Hail up to 1 3/4 inches in diameter was reported in Hot Springs, Arkansas with the passage of the cold front. At 0340 UTC 15 December an F-3 tornado touched down four miles southwest of West Memphis, Arkansas. The tornado traveled 25 miles northeastward into Tennessee, just north of Memphis. The tornado devastation resulted in six fatalities, 221 injuries, damaged 300 homes, and displaced 1,529 people.

ACTIVITY

OBJECTIVE: The development of this storm system over the mainland United States provides a good opportunity to analyze the surface structure of a record setting winter storm. The purpose of this activity is to analyze and interpret weather data plotted on a surface chart to see the evolution of the pressure and temperature fields in a winter cyclone, and to track the motion of the low center and cold front.

MATERIALS:
§ #2.5 or #3 pencil (harder lead will make analysis easier to erase),
§ #2 pencil (to finalize analysis)
§ red and blue pencils, and large pink or white erasure

PROCEDURE:
Plotted data are provided for four times during the evolution of the explosive winter storm. Additional explanation of data plots and analysis are provided in Lab 4.
1. Analysis Background
a. There is no general rule about the correct place to start. However, it is usually best to begin at a place where the pressure distribution seems apparent and the reports are ample. Use a harder lead pencil and draw lightly so that it is easy to make erasures. Even the most experienced analysts make generous use of their erasures.
b. Lines of constant pressure on a surface map are called isobars. To analyze the pressure field, think of pressure as you would altitude on a contour map of a smooth mountain. In areas where tight pressure gradients exist, a good practice is

to draw the isobars for larger intervals and then draw in the intermediate values. Fig. 6.14 provides a sample analyzed weather map for 15 December at midnight (00) Universal Time Convention (UTC)* (7 AM EST), which corresponds to the plotted map in Fig. 6.15a.

c. Isobars are smooth, continuous lines except where they cross sharply defined fronts and at the edge of the map. They should never cross or join other isobars.

e. Surface winds will cross the isobars from high to low pressure at an angle of ~10°-30°. The size of the angle depends on the magnitude of the friction from grass, trees, building, etc. The rougher the terrain the greater the angle formed between the isobar and the wind direction.

f. The speed of the wind is inversely proportional to the spacing of the isobars. Closely spaced isobars mean a stronger pressure gradient and stronger wind.

g. Plotted values may be in error due to the method by which sea-level pressure values are obtained. The error may be particularly large in mountainous sections of the country. In all cases, consider the possibility of error by the observer in reading, coding, decoding, or plotting the pressure values. Circle in red any observation found to be suspect. (Note: as discussed in Lab 4, the NWS plots sea-level pressure data in tenths of mb, and drops the leading number, so that 1011.1 mb is plotted as 111. Data in this lab are plotted to the nearest whole number, e.g., 1011.1=1011.)

h. The occurrence of thunderstorms at or near the station may cause pressures that may vary considerably from surroundings reports.

i. Lines of constant temperature are called isotherms. Isotherms can include more wiggles than isobars, reflecting the impact of local surface variations on this variable.

2. Sea-Level Pressure Analysis

a. Analyze the surface-pressure fields in Figs. 6.15a-d. Draw isobars every 8 mb (begin with an isobar value divisible evenly by 8, e.g., 1000, 1008, 1016...). Analyze quickly just to get the sense of the distribution.

b. Now adjust the isobars to fit the winds, making them smooth, with smooth

_____

gradients as well. Strive for gradual changes in the spacing between isobars. Add isobars for every 4 mb, keeping in mind that you wish to show as smooth a pattern as possible without undue disregard of the data. Analyze to the edge of the map.

c. Label the high and low-pressure centers. Use block type letters to designate location of highs (blue H) and lows (red L).

d. Label all contours. Label open isobars at both ends; they may also be labeled at other points. Label closed isobars at a point that does not overwrite plotted data. Break the isobar at that point to permit the entry of the value. For a series of closed isobars, the labels should be arranged for form an easily read line of numbers (see Fig. 6.14).

3. Surface Temperature Analysis

Analyze the temperature field in Figs. 6.10a-d. The standard contour interval for isotherms is 5° F. However, given the large range of temperatures in this storm, *draw contours every 10° F* (e.g., ...45, 55, 65°, ...etc.). Label all contours as before.

4. Finding the Cold Front

Using wind shifts, temperature contrasts and current weather as clues, locate and place the cold front in dark blue pencil on your analyzed maps (study Fig. 6.14 as an example). Use the previous frontal position as a guide to future positions; fronts tend to move in one direction.

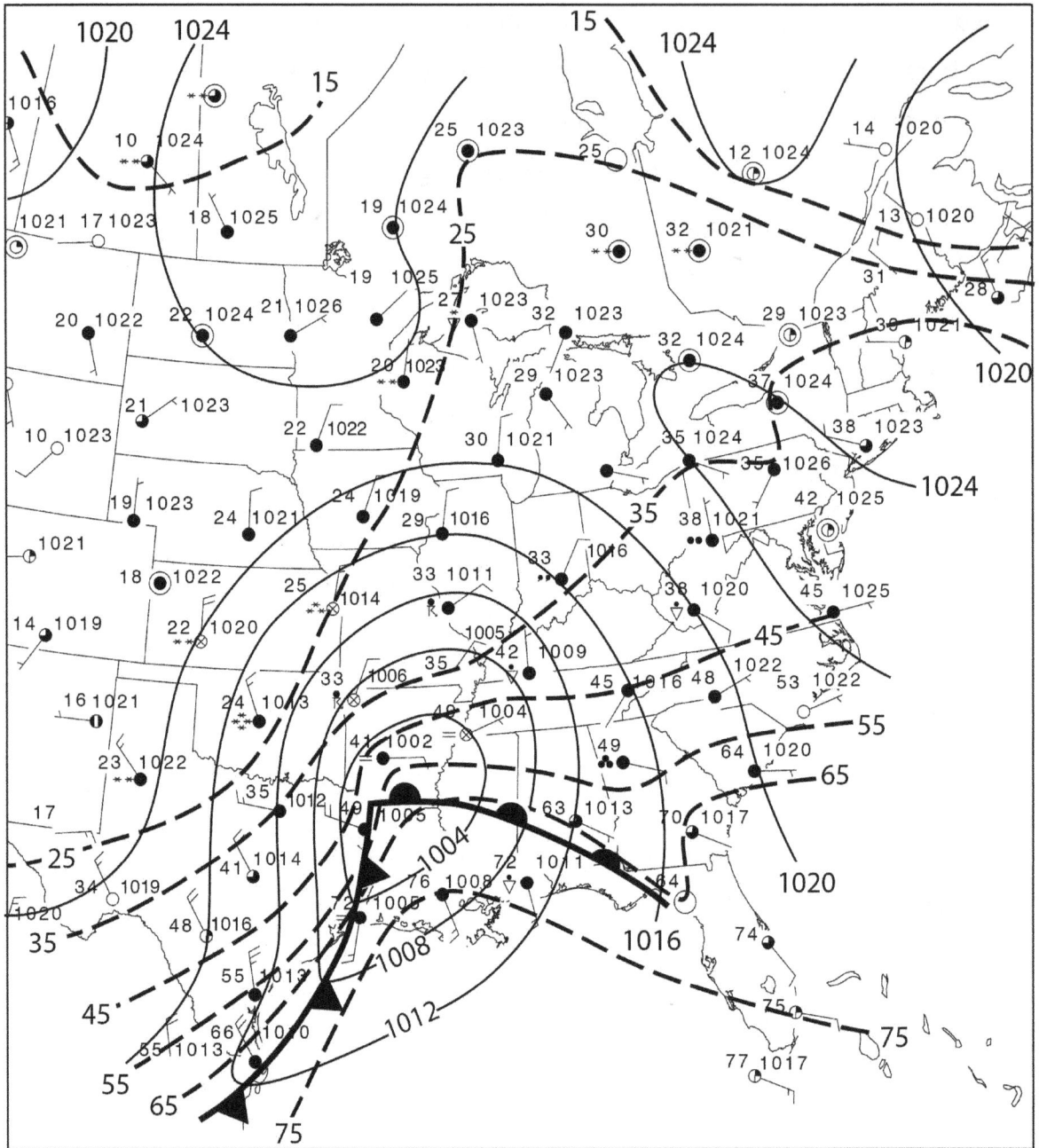

Figure 6.14 Analysis of sea level pressure (solid contours every 4 millibars) and temperature (dashed contours every 5° F) for weather observations at 00 UTC (7 AM EST) 15 December 1987.

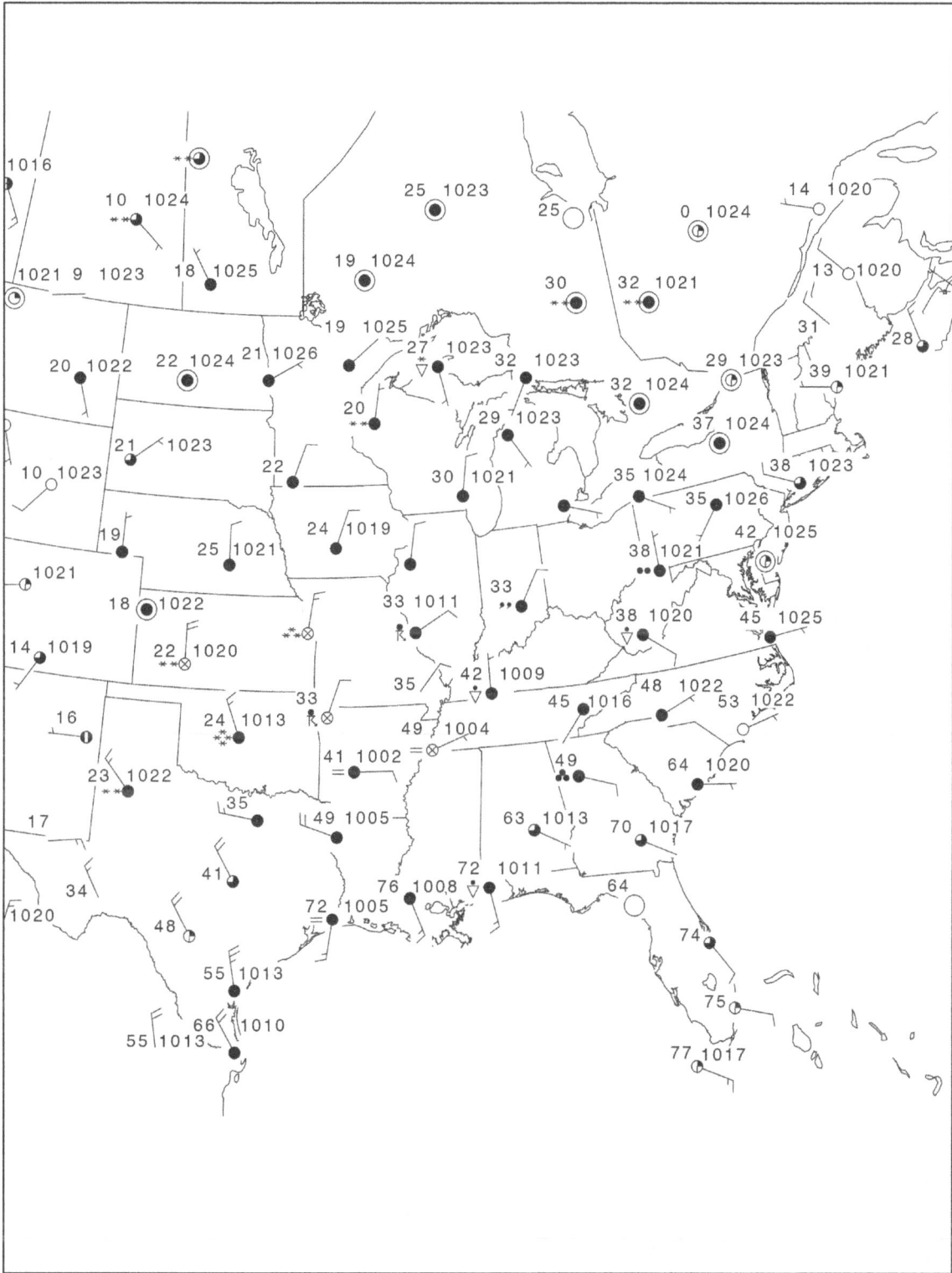

Figure 6.15a Plotted weather observations for 00 UTC 15 December 1987.

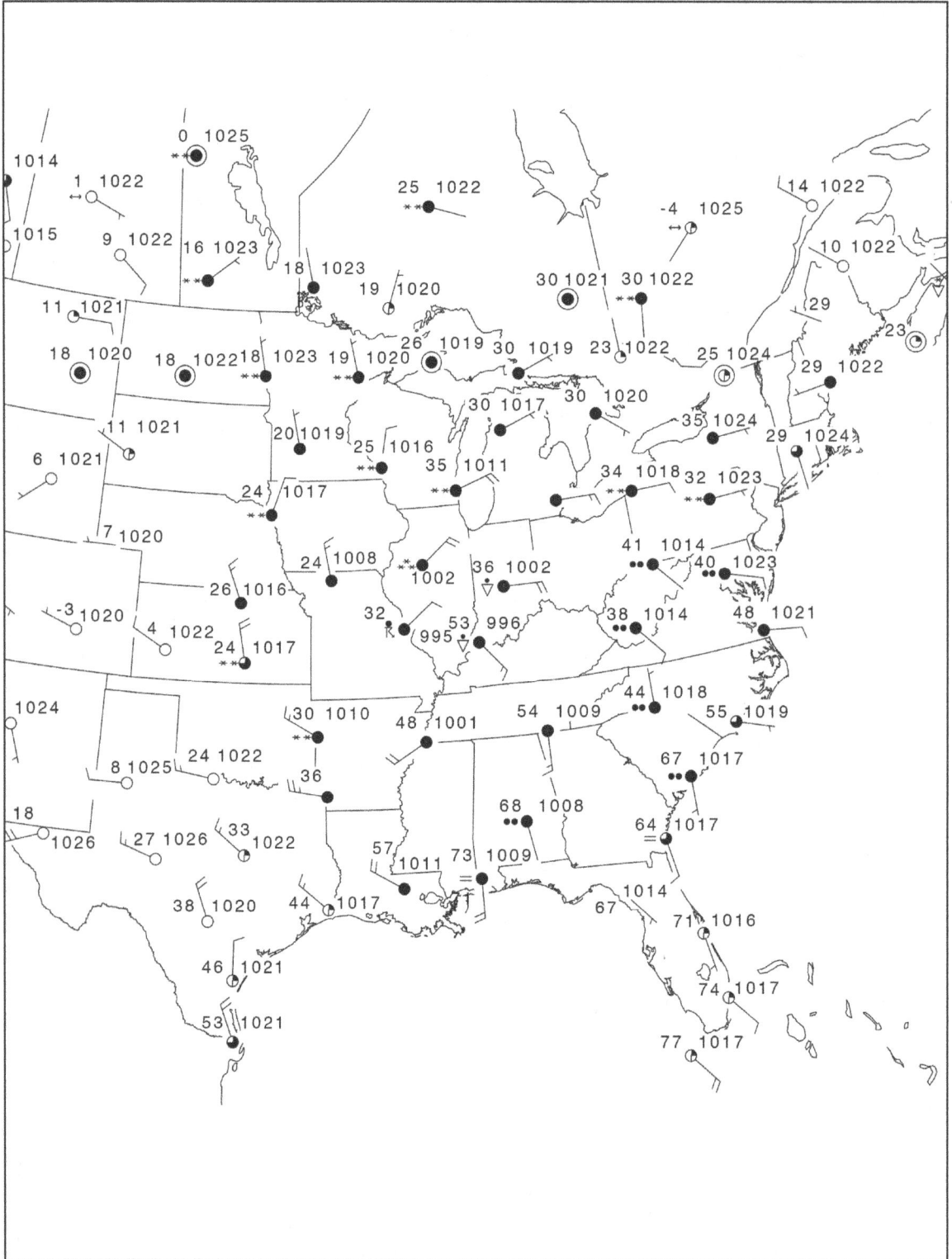

Figure 6.15b Plotted weather observations for 06 UTC 15 December 1987.

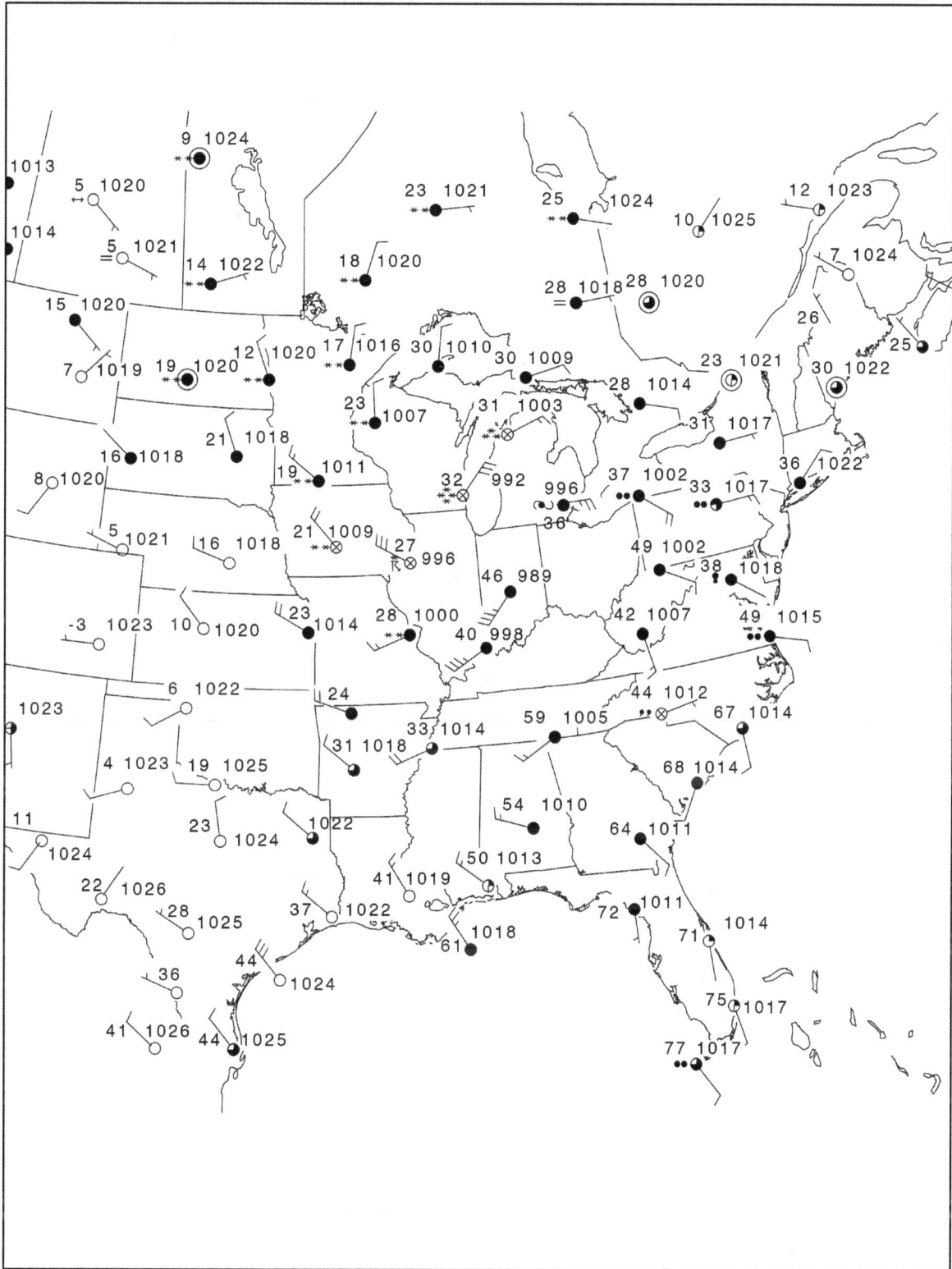

Figure 6.15c Plotted weather observations for 12 UTC 15 December 1987.

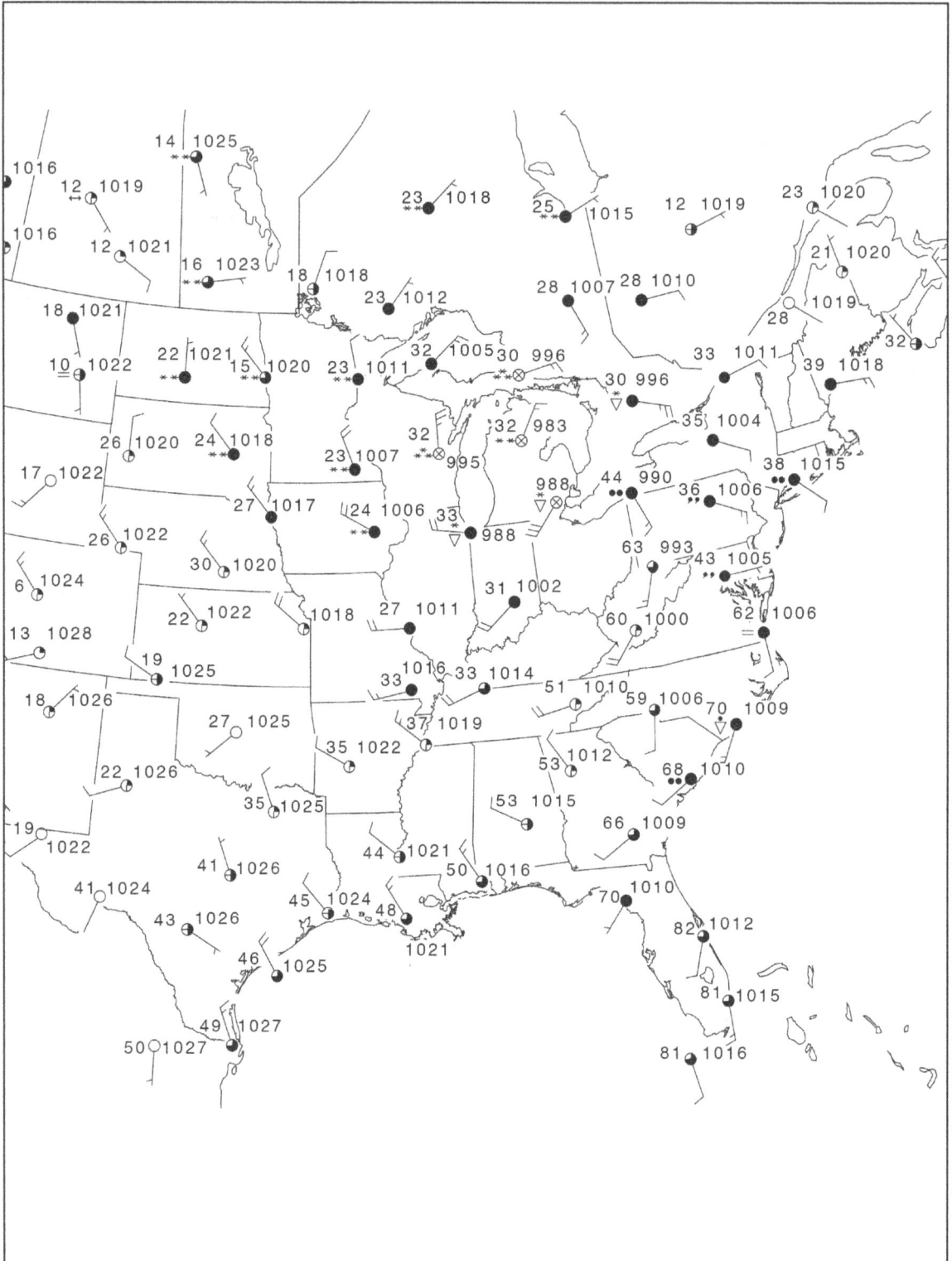

Figure 6.15d Plotted weather observations for 18 UTC 15 December 1987.

QUESTIONS:

1. In Seattle the atmospheric pressure is 1010 mb. The temperature is 54°F. The dew point temperature is 50°F. The wind speed is 15 knots, from the southeast. The cloud cover is fifty percent and it is foggy. At the time of the last observation, the pressure was 1011.5 mb. Draw the station model for Seattle below (refer to Lab 4 for station model explanation).

2. Qualitatively, what is the relationship between wind speed and spacing of isobars in your surface analyses?

3. What is the relationship between the isotherms and the position of the low-pressure center?

4. Qualitatively, what is the relationship between the isobars and the wind direction?

Lab 31: Blizzards

INTRODUCTION

"The Blizzard of '93" was a major winter storm that brought heavy snowfall along the Appalachians from Mississippi to Maine (Fig. 6.16). Electric power was lost in tens of thousands of homes and businesses due to the frozen precipitation and high winds. The death toll directly and indirectly related to the storm was estimated to be over 270 and the estimated cost of the storm is over $3 billion!

ACTIVITY

OBJECTIVES: To goal of this activity is to analyze the relationship between the storm track and the hazardous winter weather it produced.

PROCEDURE:

1.   On the accompanying map (Fig. 6.16) analyze the snowfall pattern by drawing contours of equal snow depth every ten inches (0, 10, 20, 30,..etc). Add a dashed contour for 5 inches. Allow your contours to connect stations that are not so close together when it seems reasonable to do so, given terrain enhancement of snowfall by the Appalachian Mountains.

2.   Place a bold "L" on the snowfall map for each position given in the Table 6.1 below. Label each low position with the day/time.

Table 6.1 A history of the central sea-level pressure and track of the storm.

Day/Time	Latitude	Longitude	Pressure	Remarks
12/ 7 pm	28.2 N	89.0 W	989 mb	SSE of New Orleans, LA
12/11 pm	30.0	86.2	983	SW of Pensacola, FL
13/ 7 am	32.0	83.0	973	30 mi. NW of Alma, GA
137 pm	38.7	75.8	960	30 mi. SW of Dover, DE
13/11 pm	40.9	74.3	962	20 mi. NW of LaGuardia, NY
14/ 7 am	45.0	68.1	965	NE of Bangor, ME

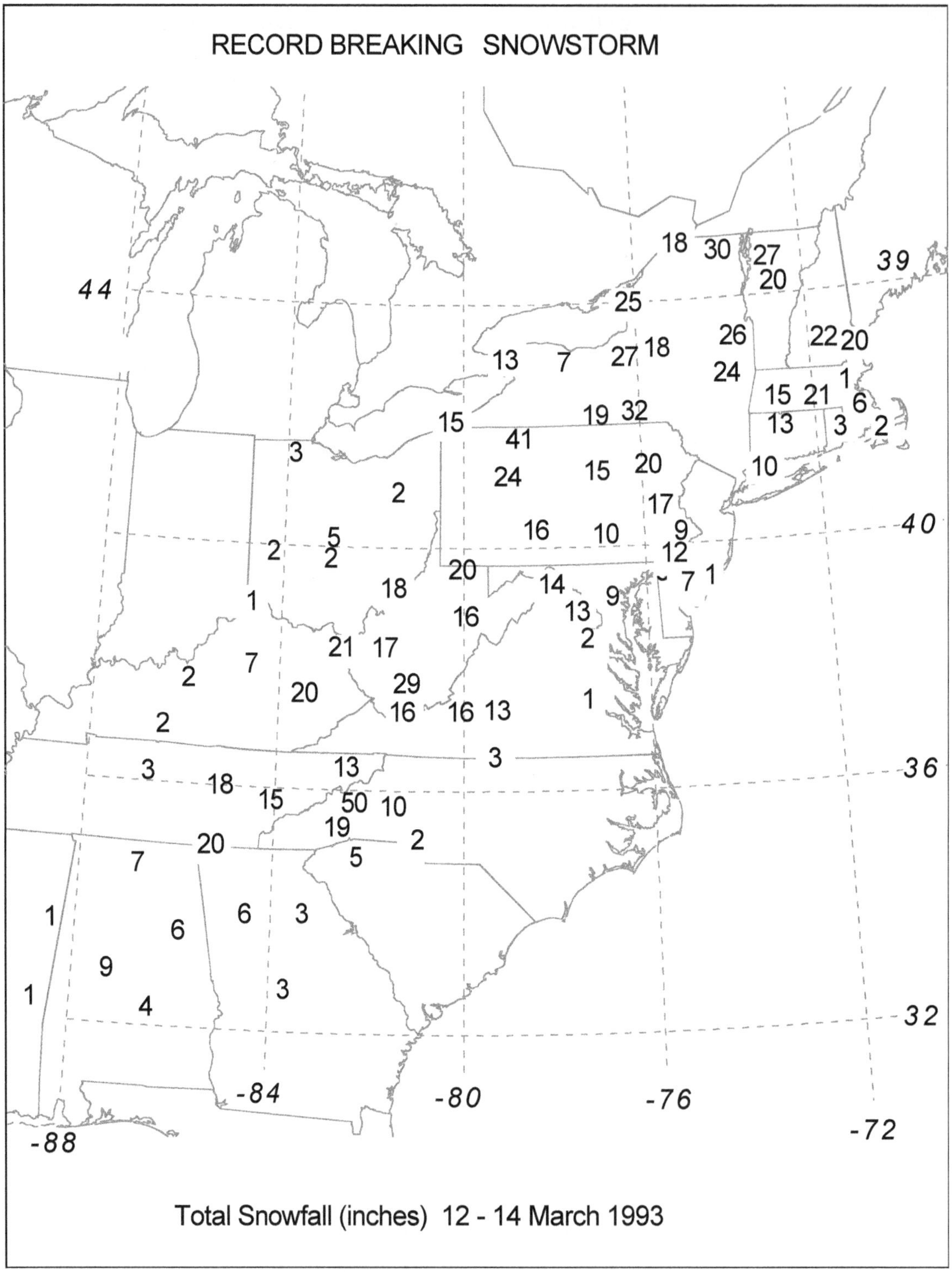

Figure 6.16 Plotted map of snowfall for severe blizzard of March 1993

QUESTIONS:

1. What is the relationship between the track of the cyclone center and the heaviest snowfall? Explain this relationship in terms of the wind circulation and frontal structure of winter storms? Why might you expect the coastal Carolinas to escape the snow in this case?

2. Note where the heaviest snowfalls occur in your analysis. What factors could account for the very high snow totals at some reporting stations?

3. Have you ever experienced a snowstorm that was not well forecast in terms of expected snow accumulations? Note the large changes in snowfall totals between locations that are rather close to each other. Could this be a factor in the difficulty forecasters have in accurately predicting snow amounts? What might account for such large variations over small distances?

Chapter 7 Weather Forecasting

'Does the flap of a butterfly's wings in Brazil set off a tornado in Texas?'
Edward Lorenz – title of a paper on predictability given to the AAAS.

A cornerstone of modern weather forecasting is the global observational network of radiosondes and surface instruments on land and on ships and buoys over the ocean. Three additional cornerstones of modern weather forecasting are (i) weather radars, (ii) weather satellites, and (iii) fast computing. Radars and satellites have been particularly important for improvements in short-term (6-12 hour) forecasting (also called *nowcasting*) for monitoring the development and motion of severe thunderstorms, fronts, and tropical cyclones. Fast computers are critical to the tasks of data collection, processing, synthesis, weather modeling, and graphical display.

Data from radiosondes and surface observations have been treated in previous chapters. The focus in this chapter is the power of satellites, radars, and numerical weather prediction models to help us better visualize and forecast the weather.

Imagery from radars and satellites, and graphical output from weather models are widely available on the web. In recent years an increasing number of universities have set up *weather servers* that provide access to real-time radar and satellite images and allow a time-lapse sequence of images to be displayed showing how weather systems develop over time. Such a site developed by the author can be found at http://weather.hawaii.edu.

Unidata, funded by the National Science Foundation, has helped pioneer Internet delivery of weather data. Unidata keeps an updated list of weather servers at http://my.unidata.ucar.edu/content/community/participatinguniversities.html.

Weather Radars

Radar (short for radio detection and ranging) was developed during World War II after it was discovered that electromagnetic waves emitted by an antenna on the ground could be reflected off an aircraft, thereby allowing the aircraft to be detected remotely. As the power and sensitivity of military radars increased, precipitation began to appear on radar screens and the weather radar was born.

Weather radars can see through clouds and observe distant rain shafts, just the way an x-ray sees through an arm to given an image of a broken bone. The radiation from the radar is reflected off millions of raindrops. The amount of scattered radiation returned to the radar (called an *echo*) is related to the number of drops and their sizes. It turns out that the rainfall rate is also related to the number and size of the raindrops, so it is not too surprising that the weather radar has become an important tool for studying storm structure, hail, and flood potential.

The weather radar most commonly scans the sky in a series of 360° cone-shaped scans, with each scan done at a higher elevation angle than the previous. As a result of the geometry of the scan and the spherical shape of the Earth, rain areas that are observed farther from the location of the radar are also higher in the atmosphere (Fig. 7.1). A rain echo seen in the 0° elevation scan at a distance of 120 km from the radar is at and elevation of ~1 km.

A radar image that shows a 360° scan of the strength of the returned signal is called reflectivity PPI (for Plan Position Indicator) (Fig. 7.2). Forecasters typically animate or *loop* a series of reflectivity images to infer the motion of weather features such as thunderstorms, frontal rainbands, and hurricane rainbands (see Lab 43). When radar echoes are found to be stationary, especially near the radar, the possibility of spurious scatter from stationary objects, such as a building or forest, is the likely cause (Fig. 7.2). These echoes are referred to as *ground clutter.*

A vertical cross section constructed from the data of all the scans taken in one compass direction (e.g., Fig. 7.1) is called an reflectivity RHI (for Range Height Indicator) and is useful for estimating the vertical extent and structure of thunderstorms and squall lines (see Lab 32).

When a thunderstorm ingests air at low levels with a high degree of rotation (or *vorticity*), the signature of this rotation can be seen in the radar reflectivity PPI as a hook-shaped echo, because sheets of rain a pulled around the by the vortex.

Modern radars, like the WSR-D88 deployed by the National Weather Service, not only detect precipitation rates but they can also measure the component of velocity of the precipitation particles in the direction parallel to the radar beam (thus, radars are a bit myopic). The ability to measure velocity comes from analyzing the phase or Doppler shift of the returned signal, and is the same principle used by police radar to gauge the speed of cars on the highway. This added capability is very important for measuring rotation in potentially tornadic thunderstorms so that more timely warnings can be issued in cases of severe weather. Also, strong gradients in wind direction and speed (wind shear) can be detected to give warnings to aircraft of dangerous *microbursts*, produced by strong downdrafts.

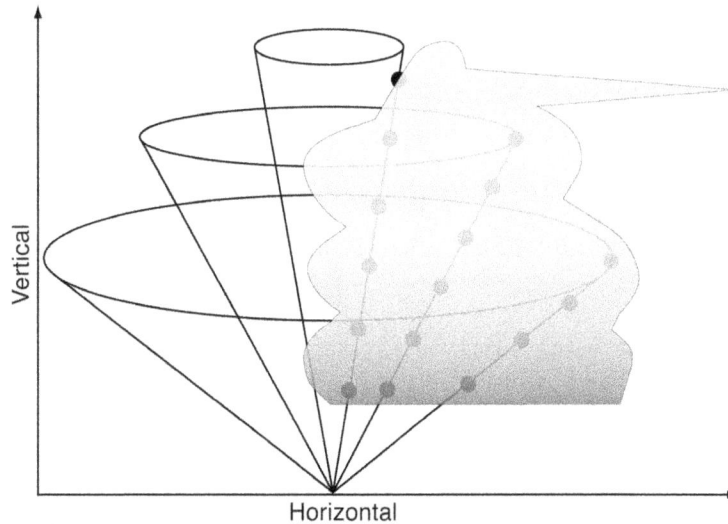

Figure 7.1 Schematic showing the conical shape of radar beam scans at three increasing elevation angles. Also shown is the construction of an RHI of a thunderstorm with data returns indicated by the dark dots along the beam in the cloud.

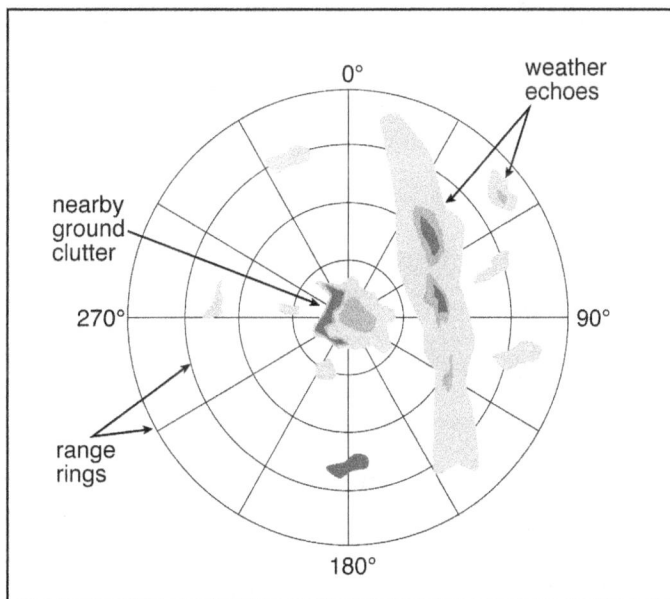

Figure 7.2 Example reflectivity PPI showing nearby ground clutter and distant echoes from rainfall. Darker shading indicates a stronger echo.

Weather Satellites

There are two primary satellite types, *polar orbiting* and *geostationary*. (Fig. 7.3) Geostationary orbits, at a height above the equator of 36,000 km (~25,000 miles), have the advantage of providing good temporal coverage for the region viewed, and data transmission is good. Movie loops or animations of evolving storm systems can be produced from geostationary imagery. Polar orbiting satellites cover a larger fraction of the globe, including polar regions not viewed by geostationary satellites, and with greater resolution since they travel in a lower orbit. However, they view only a small fraction of the globe at any given time, because of their low orbits (<1000 km). Therefore, the data from polar orbiting satellites do not have good temporal resolution and it is more difficult to receive the data transmissions.

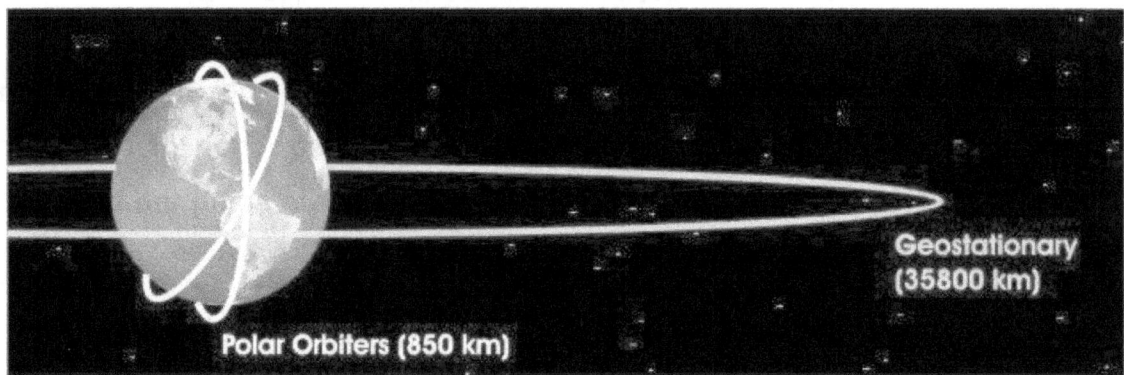

Figure 7.3 Schematic showing geostationary and polar orbits.

Meteorological satellites carry a large variety of instruments, which fall into two classes, (i) passive sensors are those that only receive energy from the Earth and atmosphere, and (ii) active sensors are those that emit radiant energy, which is scattered back toward the instrument's sensor by the Earth and/or atmosphere.

Visible (reflectivity), Infrared (temperature), water vapor channel (moisture)

Passive sensors are the instruments most commonly used to obtain the variety of colorful imagery of the Earth and atmosphere used by weather forecasters and the media. Three types of satellite imagery are most common, (i) visible (~0.4 micron), (ii) infrared (~10.7micron), and (iii) water vapor channel (~6.2□micron) at the wavelength of strongest emission by vapor molecules in the atmosphere. Each of these channels on the satellite sensors is sensitive to electromagnetic energy at a particular range of frequencies. Therefore each type provides a different view of the Earth, atmosphere,

184

and oceans. Weather observers and forecasters rely on all three types of data, and often use them together to better understand the interactions between the atmosphere, oceans, and the Earth's surface. Furthermore, satellite meteorologists can assign (or map) various shades of gray to each brightness or temperature range to allow different information to be highlighted in the final satellite image (Figs. 7.4 and 7.5). When color is used in addition to black and white, the resulting image is referred to as *color enhanced.* Enhanced imagery is a method meteorologists use to aid them with satellite interpretation. The enhancement enables a meteorologist to easily and quickly see features that are of special interest to them (Fig. 7.4). Usually they look for high clouds or areas with a large amount of water vapor. In an infrared (IR) image cold clouds are high clouds, so the colors typically highlight the colder regions (Fig. 7.5b). In a water vapor image, white areas indicate moisture and dark areas indicate little or no moisture, so the colors typically highlight areas with large amounts of moisture (Fig. 7.5c)

Visible imagery has the highest resolution (Fig. 7.5a). For geostationary satellite imagery the resolution is typically ~1 km for visible imagery and several km for IR and water vapor imagery. Visible imagery also can be used to help distinguish between a thin high cloud (cirrus) and a deep convective cloud (cumulonimbus).

IR emissions from cloud top can be converted to an equivalent emission temperature (Planck's law), and a temperature sounding can then be used too give height (see Figs. 7.4 and 7.5b). Similarly, sea-surface temperatures can be obtained in cloud-free regions.

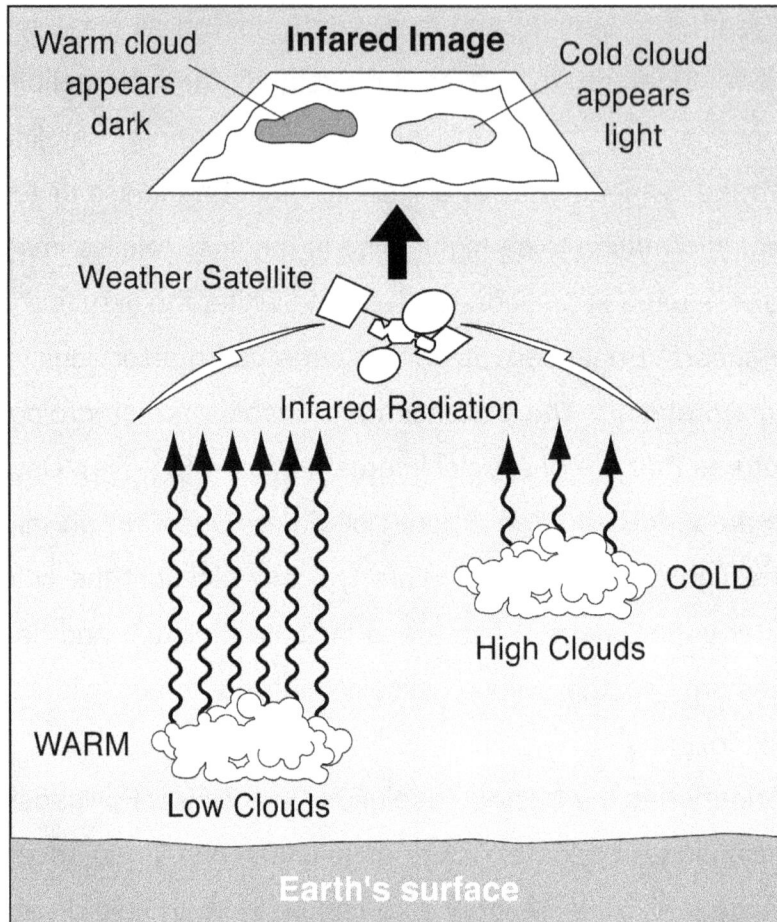

Figure 7.4   Schematic showing the imaging of infrared (IR) radiation from clouds.

The median level of emission by water vapor in the atmosphere viewed in the water-vapor channel is ~400 mb. Therefore, by observing the IR emission of the water vapor, circulations in the upper troposphere are revealed (Fig. 7.5c). Sinking motions in the upper atmosphere result in dark areas, while rising motions produce clouds and various shades of gray and white (or colors), depending on the enhancement used.

Wind velocity across the globe is retrieved through automated analysis of successive images from geostationary satellites to track cloud features. IR temperatures of the clouds are compared to available sounding or model data to estimate the height of the inferred winds (Fig. 7.5d).

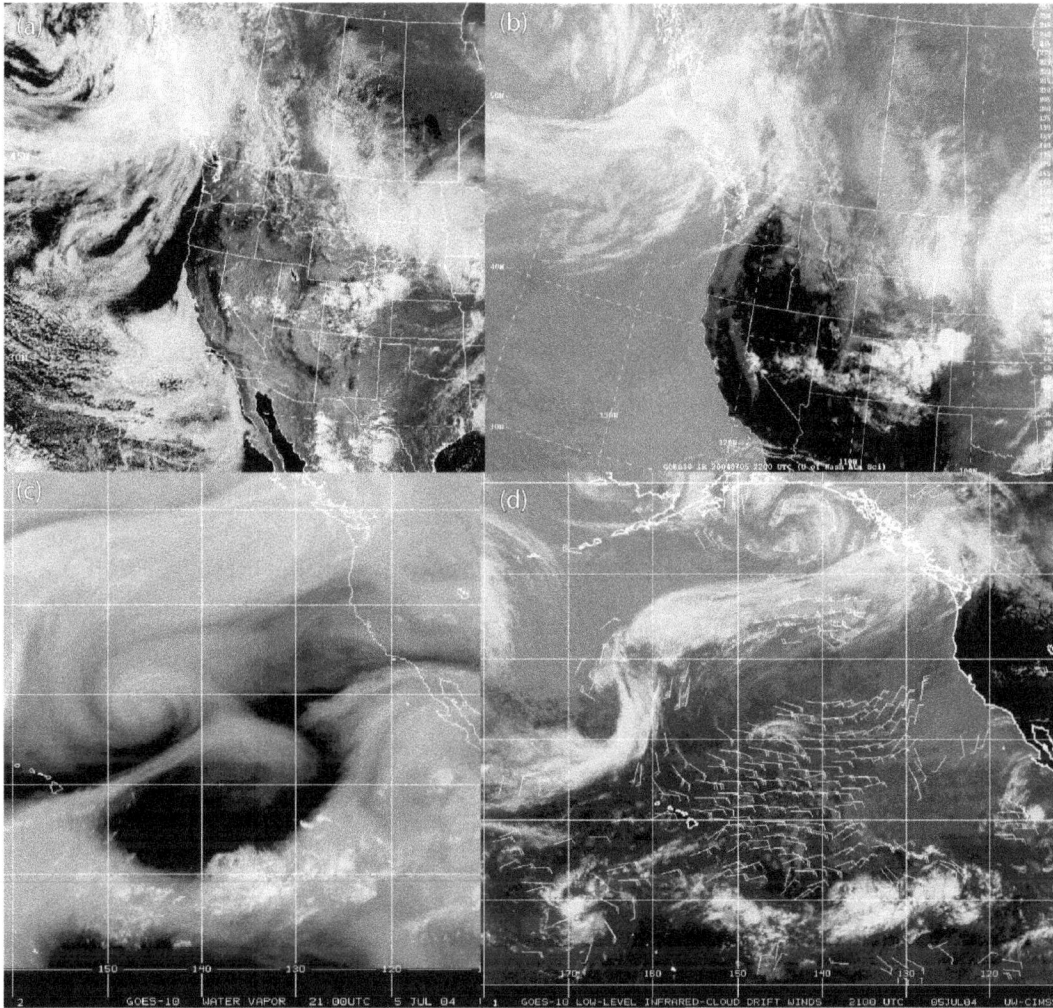

Figure 7.5 GOES12 imagery for 1900 UTC 5 July 2004. (a) visible, (b) enhanced infrared, (c) enhanced water vapor, and (d) infrared with cloud drift winds.

Active instruments emit radiant energy in a variety of wavelengths to obtain a variety of observations. Satellite active sensors include lidar (visible and UV), radar (microwave 3-5 cm), and GPS signals (microwave 19 and 22 cm). Active sensors tend to be heavy and large power consumers. Therefore, their deployment is generally limited to lower, polar or inclined orbits. GPS satellites are placed in inclined (55°) orbits at 20,000 km. Data sets collected by active instruments include the following.

1) Profiles of wind speed are obtained using the Doppler shift in returned UV lidar signal scattered by air molecules in atmosphere down to cloud top or aerosol layer.

2) Rainfall rates are estimated using the returned power of microwave signal scattered by cloud and precipitation particles.

3) Surface winds are estimated over the ocean by analyzing the spectral reflectance (which depends on the size of the waves) of microwave (radar) energy returned to the sensor from the ocean surface.

4) Temperature and moisture profiles can be obtained through observations of the delay and bending (refraction) of GPS signals by the atmosphere as a GPS satellite sets with respect to a space-based GPS receiver.

Interpreting Satellite Imagery

i) Analyzing cloud patterns – In general, the clouds shown in satellite imagery can be classified as layer clouds or convective clouds. Layer clouds tend to cover large areas and are indicated on a satellite picture by an area of uniform brightness or temperature (Fig. 7.5). This type of cloud is formed by large-scale rising motion in the atmosphere that is often associated with lows or fronts. Convective clouds are usually formed by air being heated from below and are especially common in the tropics and over the oceans behind cold fronts. The rising air generates cloud with the surrounding descending air being cloud-free. The individual clouds can be identified on a satellite image, and it is possible to look at the build-up of thunderstorm cells, especially over continents during spring and summer afternoons.

ii) Identifying the location and evolution of winter storms (low surface pressure)
Satellite images are particularly helpful in locating depressions and fronts. A series of images can be used to illustrate the evolution of winter storms. The cloud shield associated with the developing storm often starts out as a leaf shape in IR imagery, and evolve into a comma shape, and then a spiral as the storm matures and occludes. Depressions or low pressure centers can be picked out by their distinctive swirl of cloud, and frontal systems can often be seen as a wishbone-shaped area of cloud radiating from a depression. A cold front is often clearly shown as a distinctive trailing edge of the left-hand prong of the wishbone pattern. With experience, meteorologists can mark the location of fronts on satellite images, particularly cold fronts, with

reasonable accuracy. This ability is particularly useful over the ocean where ship and buoy data are limited.

iii) Inferring the location of anticyclones (high pressure) – In anticyclones, the air is descending and warming - this means that thick cloud will not form, so areas of high pressure, especially blocking anticyclones, can be easily identified by the relative absence of cloud and the ground and coastline can be clearly seen in visible and IR images. Some web sites will overlay the analysis and satellite data to provide a combined image.

(see for example http://www.atmos.washington.edu/~ovens/loops/)

iv) Estimating wind speeds and the movement of frontal systems – It is possible to calculate how far individual cloud features have moved in successive images and estimate their velocity. The height of the cloud feature is estimated by comparing it IR cloud top temperature with near by temperature soundings from radiosonde or aircraft. Similarly, the motion of cloud bands associated with cold and warm fronts can give an estimate of the motion of the fronts.

v) Studying global pressure belts – In global or hemispheric images the location of high- and low-pressure belts at the planetary scale can be distinguished. Larger features such as the intertropical convergence zone (ITCZ) can be identified over the Pacific Ocean (see lower part of Fig. 7.5d). By looking at images from different times of the year, it is possible to see how these belts shift. Images over the Indian Ocean can be used to study the build-up of the Indian Monsoon.

vi) Analyzing daily temperature changes in IR images

A series of daily IR images can help show diurnal temperature variations as well the contrasts between land and sea temperatures.

vii) Tracking the movement of tropical storms – Satellite imagery are particularly important for tracking tropical cyclones and for gauging their size and intensity in the absence of aircraft data. Forecasters rely on loops or sequences of satellite images to track the progress of hurricanes.

Weather Forecasting

There are a number of approaches that people have used historically to construct a weather forecast, even without the aid of a computer. These include the following methods.

Persistence forecast – this is simply a prediction that future weather will be the same as the present weather. For example, if it is raining now, a persistence forecast would say that the rain will continue. This method of forecasting is most accurate for short time periods and becomes less and less accurate with increasing time.

Steady-state or **trend forecast** (**nowcasting**) - this method assumes that weather systems that are moving at a given velocity, will continue moving at that velocity. This method is fine for systems that are not accelerating and not amplifying. It also works best in the shorter time ranges.

Analogue method- this method assumes that the forecaster can find a reasonably similar weather pattern in the past and use the sequence of events that occurred at that time to predict the sequence of events that will occur in the future. Experienced forecasters use this method implicitly as they recognize patterns and situations that they have observed in the past.

Climatological forecast- is a forecast based on the climatology of a particular region and is sometimes expressed in terms of a probability. For example, if Minnesota had 27 white Christmases during a 30-year period, then the chance of a white Christmas in Minnesota is ~90%.

Combination forecast – applies intuition gained from the above approaches with guidance provided by numerical weather prediction models. As the validation time for the forecast extends farther into the future, guidance from numerical weather prediction models becomes increasingly important. The forecaster then makes a prediction based on the guidance, a practical interpretation of the weather situation, and local geographic features that influence the weather within the specific forecast area.

Numerical Weather Prediction (NWP)

Numerical weather prediction or weather modeling currently employs the world's most powerful super computers to solve the equations of motion forward in time and produce maps of the future state of the atmosphere. The process of producing a numerical forecast of the future state of the atmosphere is complex, and can be outlined as follows.

(i) The initial conditions for the model simulation are provided by surface and upper air observations taken twice daily (0 and 12Z).

(ii) The observations are interpolated to grid points and modify a first guess field for those grid points put out by the previous model run.

(iii) The laws of physics[3] for the atmosphere are applied to the initial conditions in the model and are integrated forward in time to calculate values at each grid point in the future.

(iv) Model output (*prognostic*) charts and model output statistics (MOS) are then made available to the forecaster as *guidance* to construct specific forecasts.

The accuracy of numerical weather prediction is inherently limited by the natural chaos that characterizes the atmosphere. In addition, model accuracy is reduced by bad data, simple representations of complex physical processes[4], and poor resolution in both the data and grid spacing. If the grid spacing is reduced by half the computational expense is increased by a factor of 2^4 or 16. Moreover, effective use of higher resolution in the model requires higher resolution in the observations. Assimilation of satellite and radar data provide some of that needed resolution in modern NWP models.

The model doesn't forecast many of the variables we are interested in at the levels we want. For example, the model does not predict a high temperature; it does not predict a ceiling height or many of the variables that we are interested in knowing. The actual numbers that you see on the maps and on the MOS output are produced

---

[3] Conservation of motion (momentum), conservation of mass, conservation of heat (thermodynamic energy), and conservation of water (mixing ratio/specific humidity, condensate).

[4] Including, grid-scale precipitation (large scale condensation), deep and shallow convection, microphysics in clouds, turbulence and heat exchange at the surface, radiation, cloud-radiation interaction, diffusion.

from statistics that use both observed variables and model predicted variables as predictors.

Good forecasters use the model forecasts only as guidance. They know the strength and weaknesses of the different models and their output. They know where the problems arise in model forecasts and recognize those situations where the models will have trouble. To do this, the forecaster looks carefully at analyses of the observations and scrutinizes all the models. The forecaster looks for bad data in the analyses, he recognizes where a parameterization in the model will results in a bad forecast, and he will look for the position of features in the analyses that may not be correctly analyzed by the model because it's located in a data poor area. He realizes the weaknesses of MOS. MOS is usually incorrect when the model forecast is bad. It is also bad when the pattern is anomalous (not enough of a statistical base). Still another time MOS misses the mark is during the change of seasons. The equations for the different variables forecast by MOS change at these times so that there really isn't a good set of equations that are statistically valid during the changeover.

Lab 32: Weather Radar's X-Ray Vision

INTRODUCTION

Since their first use for detecting rain bands in World War II, weather radars have become one of the most important tools in weather forecasters' arsenal. The strength of the signal reflected by precipitation particles gives information on the rate of rainfall and on the structure of storm systems. Strong echoes from high altitudes in thunderstorms are an indication of hail. A bounded weak-echo region (BWER) observed in a reflectivity RHI (vertical cross section) of a thunderstorm is indicative of a very strong updraft, and represents another distinctive feature seen in severe thunderstorms.

ACTIVITY

OBJECTIVE: The purpose of this lab is to understand radar reflectivity data and its utility in observing and forecasting the weather.

PROCEDURE:

1. Identify potentially severe thunderstorms in Fig. 7.6 and circle them in red.

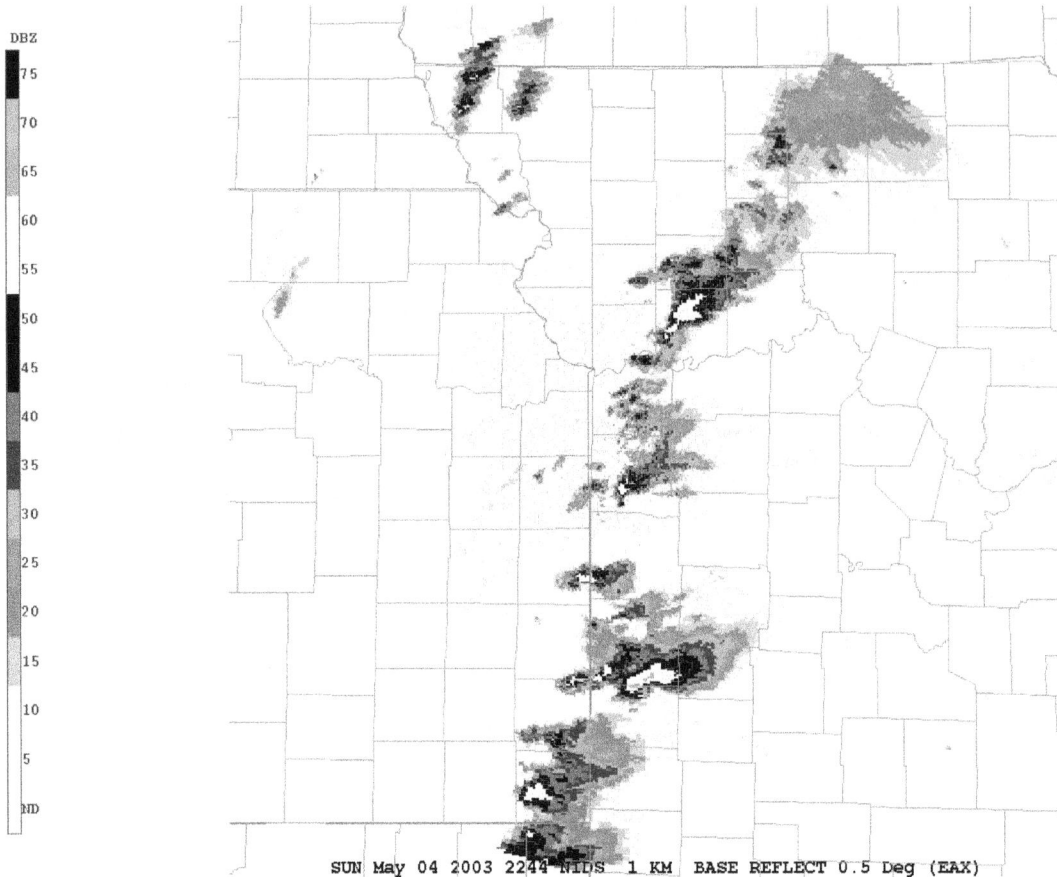

Figure 7.6 Radar reflectivity PPI with echoes over eastern Kansas on 4 May 2004.

2. A vertical cross section (RHI) of one of the storms in Fig. 7.6 is provided in Fig. 7.7, showing the effective radar reflectivity factor in decibel units (dBZ_e). Z_e is in units of mm^6/m^3. Each number shown represents the mean Z_e for a region of dimensions 1.5 km in range and about 1□ in azimuth centered on the point where it is plotted. Assume that beam filling is complete and that there is no attenuation. The freezing level on this day was 5 km (~18,000') AGL.

3. Draw contours for values of 10, 20, 30,....., 70 dBZ_e. Use colored pencils to bring out the patterns.

4. Find and label the bounded weak echo region (BWER).

5. Use the data in Table 7.1 to make a graph of the rainfall rate at the surface as a

function of distance from the radar (10 to 60 km). Indicate on your graph regions of convective versus regions of stratiform rainfall.

Table 7.1 Standard WSR-88D relationship between reflectivity (Z) and rainfall rate (in/hr). Reflectivity ≥ 50 over the continent is often associated with large hail.

Z	inches/hr
20	0.02
25	0.04
30	0.09
35	0.21
40	0.48
45	1.10
50	2.50
53	4.09
55	5.68
57	7.89
60	12.92

QUESTIONS

1. What features in Fig. 7.7 indicate that this is a potentially severe thunderstorm?

2. What does the presence of a BWER imply about the storm?

3. What clues are there in the data that suggest that this may be a severe hail storm?

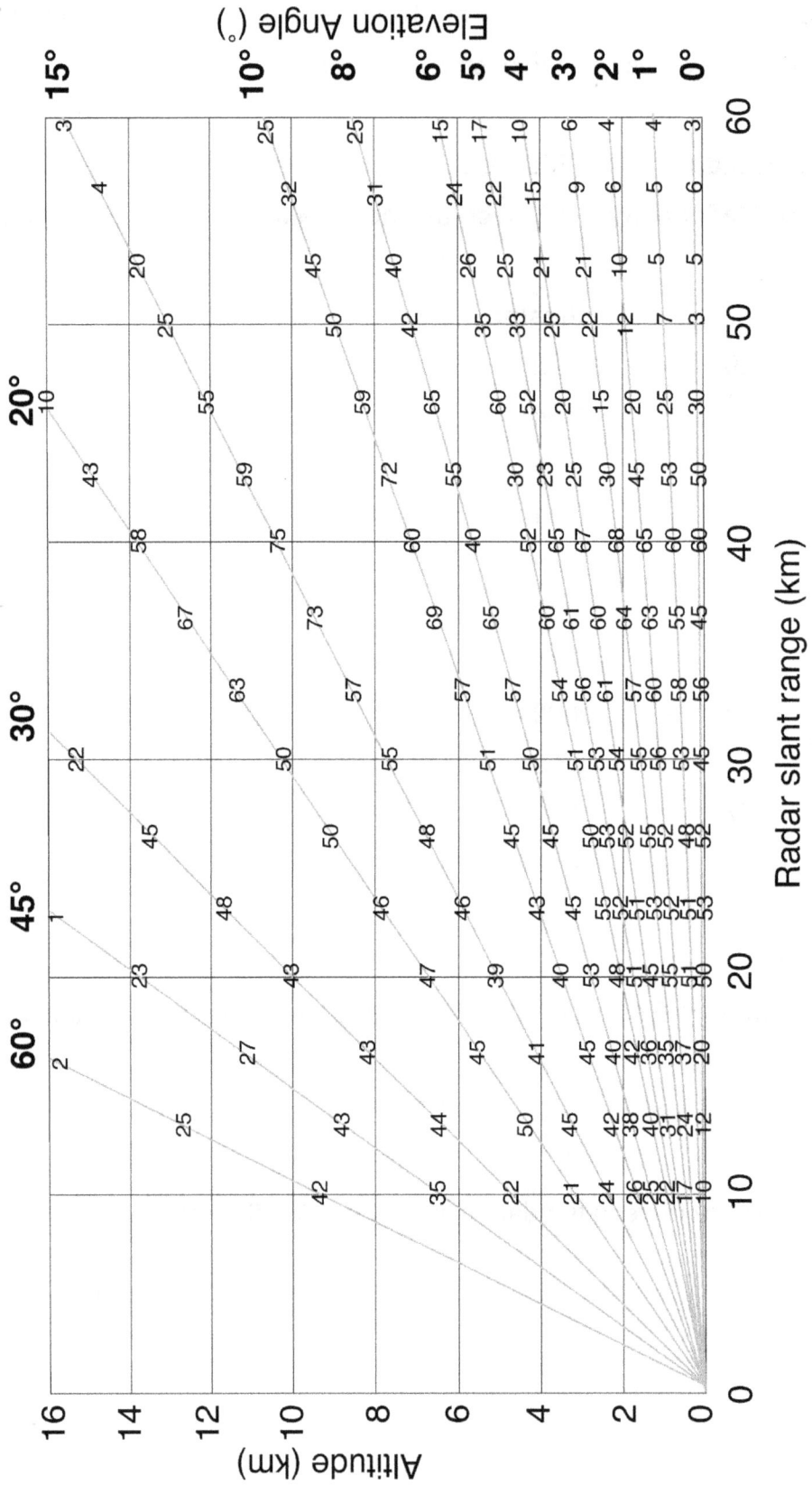

Figure 7.7 Plotted reflectivity data along an RHI.

Lab 33: Doppler Radar can See the Wind

INTRODUCTION

Doppler radars can measure the wind speed by sensing the change in phase or frequency of the signal scattered back to the radar by moving raindrops. By carefully measuring the frequency shift, the wind speed component along the direction of the radar beam can be measured. As noted in the introduction to this chapter, weather radars scan the sky in cone shaped sweeps (Fig. 7.1) and can only see the wind component along the beam, toward or away from the radar. Thus, radars suffer a bit from tunnel vision. As the radar beam sweeps around it will intersect places where the wind direction is exactly perpendicular to the radar beam. Since the radar can only detect motion in the direction of the radar beam, it will register no wind velocity at this location. The line along which the wind direction is perpendicular to the radar beam is called the *gray line*, because it was shown as gray in early radar PPIs (Fig. 7.7). The gray line usually appears as a white line on modern displays of radar PPI. The gray line in a velocity PPI provides a convenient means for determining the wind direction. If the wind direction changes with height, the gray line will curve away from the radar.

Similarly, as the radar sweeps around it will at some azimuth angle point exactly in the direction of the wind and will record a maximum wind toward the radar in that direction. When the radar sweeps out 180° from this direction it will see a maximum in flow away from the radar (see "J" in Fig. 7.7).

If the speed and direction are reasonably steady across the circle swept out by the radar, and sufficient return is available from rain and/or aerosols across the entire scan, then both a wind direction and a wind speed can be assigned to each altitude observed by the radar and a wind profile can be constructed. Changes in wind direction with height give information to forecasters regarding advection of warm and cold air above the radar, which cause important changes in the stability of the atmosphere. When the wind turns clockwise with height, warm-air advection is taking place, and this is called *veering* of the wind with height. When the wind turns counterclockwise with height, cold-air advection is taking place, and this is called *backing* of the wind with height. The reason for this is related to thermal wind balance and is beyond the scope of this book. Warm-air advection at low levels, surmounted by cold-air advection aloft, results in a reduction in atmospheric stability, and may lead to severe thunderstorms.

ACTIVITY

OBJECTIVE: The purpose of this lab is to understand radar wind velocity data and its utility in observing and forecasting the weather.

PROCEDURE:
1. Fig. 7.8 contains two radar velocity PPIs. On each panel sketch isotachs every 20 kt for a low-level jet pattern with the jet core and maximum wind speed of 70 kt. Label the cores of the jet with a 'J'. Assume good radar returns across the entire scan.

2. On each panel of Fig. 7.8, plot wind barbs (speed & direction), using the station model convention for plotting winds, at each place where the gray line crosses a range ring. The wind direction is perpendicular to the radar beam where the beam crosses the gray line. From that point, follow a range ring around to the maximum speed indicated by your isotach analysis and that is the wind speed to be plotted on the grayline.

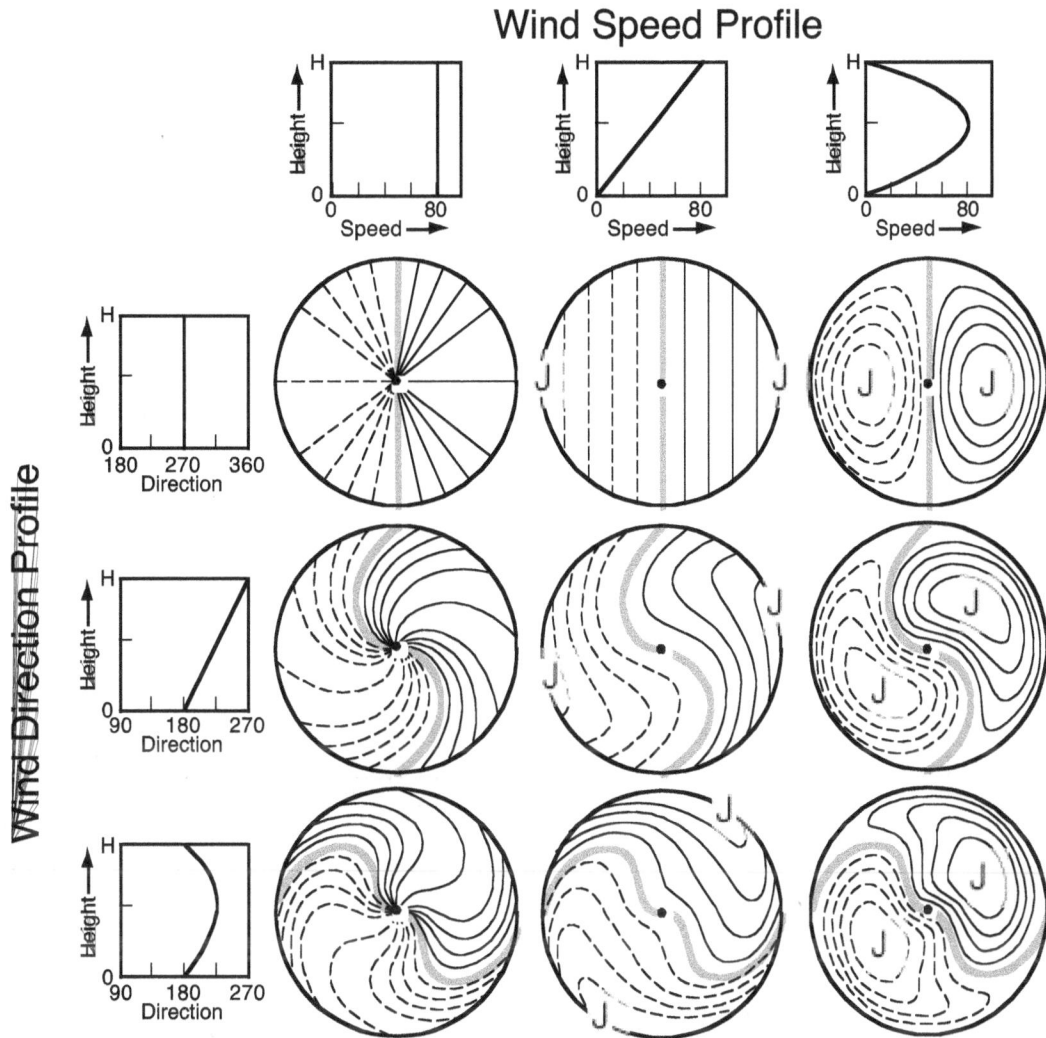

Figure 7.7 Schematic depiction of radar velocity PPIs for changing wind conditions with height in the troposphere. The thick gray line indicates zero observed wind speed in the direction of the radar beam. Changing wind direction with height is plotted on the left hand side and changing wind speed with height is plotted along the top. Resulting patterns of radar velocity (isotachs every 20 kt) are shown in the scan circles. Solid contours indicate flow away from the radar. Dashed contours indicate flow towards the radar. "J" indicates the location of a wind maximum or jet. "H" is the altitude of the radar beam at the outer edge of the data scan. The value of H will depend on the elevation angle and the range of the radar data. The middle column of circles depicts a jet aloft, whereas the right-hand column depicts a low-level jet.

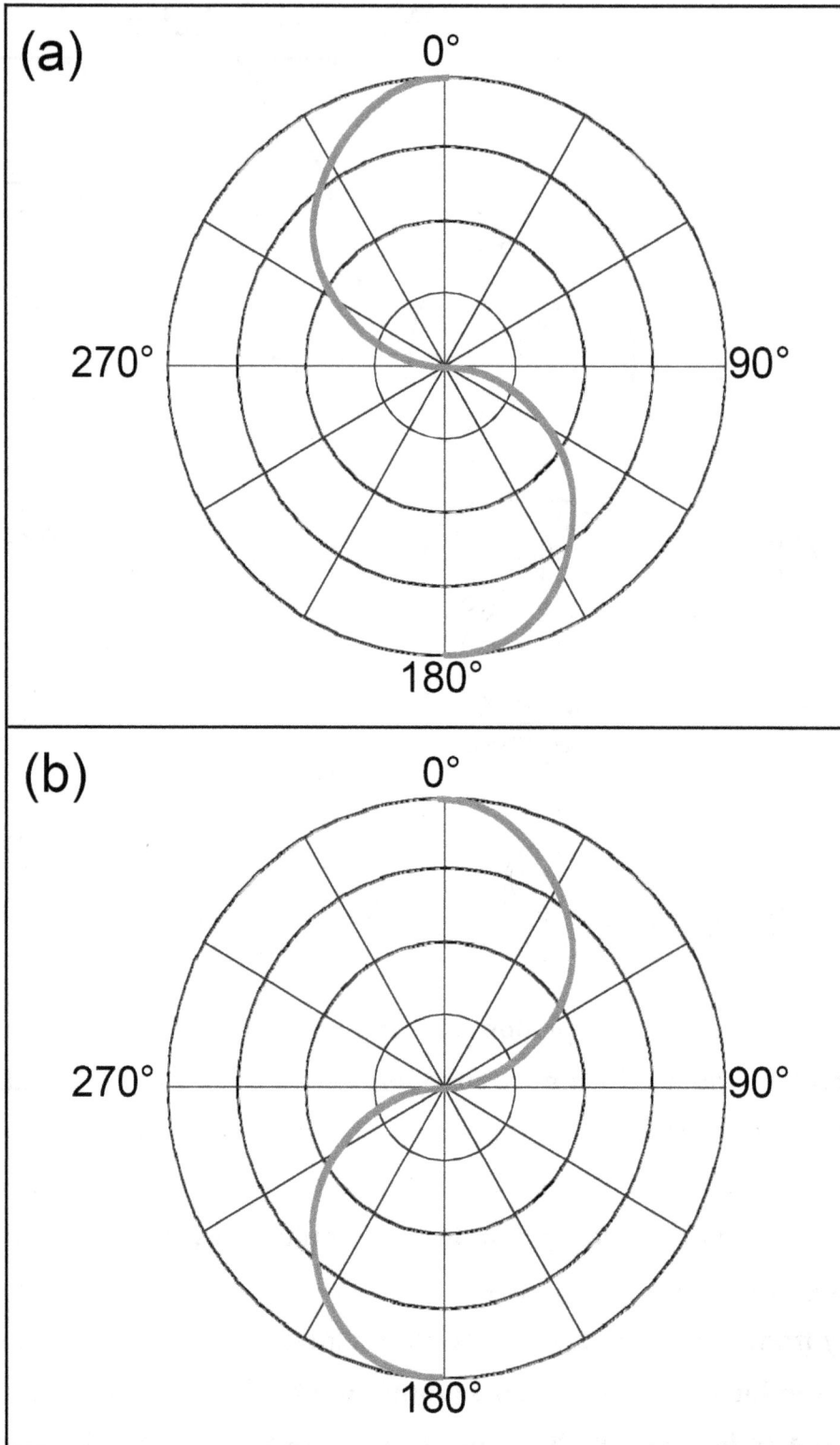

Figure 7.8 Radar PPI plot with 20 km range rings and the position of the *gray line* shown.

QUESTIONS

1. What is the wind direction at the radar site in each case in Fig. 7.8? There are two possible correct answers here, why? Choose the one that is consistent with your analysis on Fig. 7.8 in the procedure section.

2. What is the wind direction and wind speed at the jet core level for each of your analyses in Fig. 7.8?

3. Is cold advection or warm advection occurring in each case in Fig. 7.8?

4. At the level of the jet core, in what direction is the low pressure located? Assume geostrophic balance.

5. If the impact of friction is strong within the 20 km of the radar (1st range ring), what affect would that have on the shape of the gray line. Hint: recall that friction slows the wind and causes the wind direction to turn towards lower pressure (counterclockwise in the NH). Also, note that friction decreases with height.

6. What would the signature of a strong vortex (mesocyclone) look like on a radar velocity PPI?

Lab 34: Satellites – Our Eyes in Space

INTRODUCTION

Satellite imagery provides a *real-time* view of weather systems and plays a critical role in weather forecasting. Learning to understand and interpret satellite imagery is the key to unlocking their usefulness.

In a visible image space is black, and clouds are bright white, except for thin cirrus. Most water is dark, although sun glint can brighten it considerably. Land varies in brightness depending on snow cover, vegetation, soil moisture, etc.

In an infrared image, space is white (cold). High/deep clouds are white because temperature decreases in the troposphere. Low clouds are gray, often indistinguishable from the surface. The surface (land or water) varies in brightness depending on skin temperature. Sea surface temperature variations can be seen in some IR imagery.

In a water vapor image, space is white (cold). There is considerable variation in brightness in clear air depending on mid- to upper-level moisture and/or temperature. Upper-level lows (swirls of dark and light grey) and jet streams (dark streaks bounded by light grey to the south) are apparent in water vapor imagery to the trained eye. High/deep clouds are white (essentially noise to the sensor). Low clouds are not visible because they are obscured by radiation from water vapor molecules higher in the atmosphere.

ACTIVITY

OBJECTIVE: To learn to identify common types of weather satellite imagery and the types of weather features that are imaged in each.

PART I

PROCEDURE: Refer to the satellite images in Fig. 7.9 to answer the following questions.

1. What type of satellite image is shown in Fig. 7.9a; what features distinguish it as such? Indicate with arrows the axes of any upper-level jets. Label with an "L" the location of any upper-level lows.

2. Describe the main features visible in Fig. 7.9b. What type of satellite image is this; what features distinguish it as such?

Fig. 7.9a GOES8 satellite image at 1945 UTC on 28 August 1996.

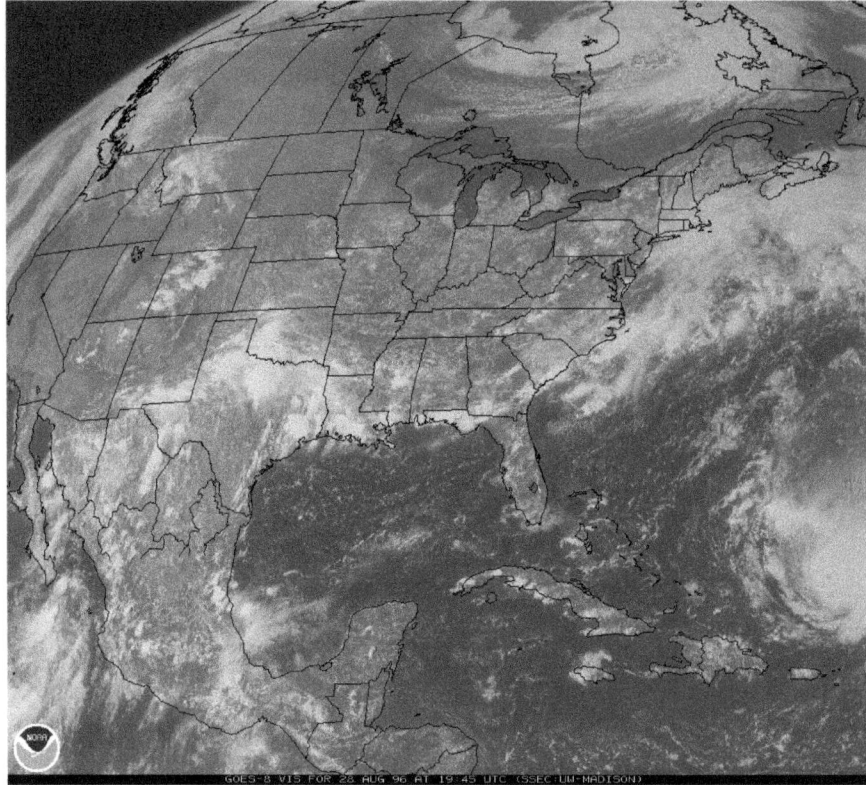

Fig. 7.9b GOES8 satellite image at 1945 UTC on 28 August 1996.

Fig. 7.9c GOES8 satellite image at 1945 UTC on 28 August 1996.

3. Describe the main features visible in Fig. 7.9c. What type of satellite image is this; what features distinguish it as such?

4. Compare the three images in Fig. 7.9 and describe what is uniquely visible in each image. In other words look for features that can only be discerned in one of the images, but not in all. Explain why this is so, based on the wavelength of the radiation being imaged.

Lab 35: Satellite Analysis of a Winter Storm

INTRODUCTION

Locating fronts and jets streams from satellite imagery is particularly useful over the oceans where surface observations may be scarce. In mature winter storms, the cold front generally starts on the warm-air side of the cloud band at its southern end and crosses the cloud band to the west side as one travels north along the front (Fig. 7.10). The cold front becomes an occluded front once the warm front is crossed. The location of the warm front is harder to pick out, and surface observations should be consulted, but it tends to be located were cloud streaks extend farther eastward. The low center is located at the center of the spiral created by the low cloud to the northwest of the cold front. The polar jet axis is generally marked by a slot of dry air on the west side of the cold front and the northern boundary of high cloud on the east side of the cold front. The direction of storm motion can be estimated by the direction of the dry slot associated with the jet axis.

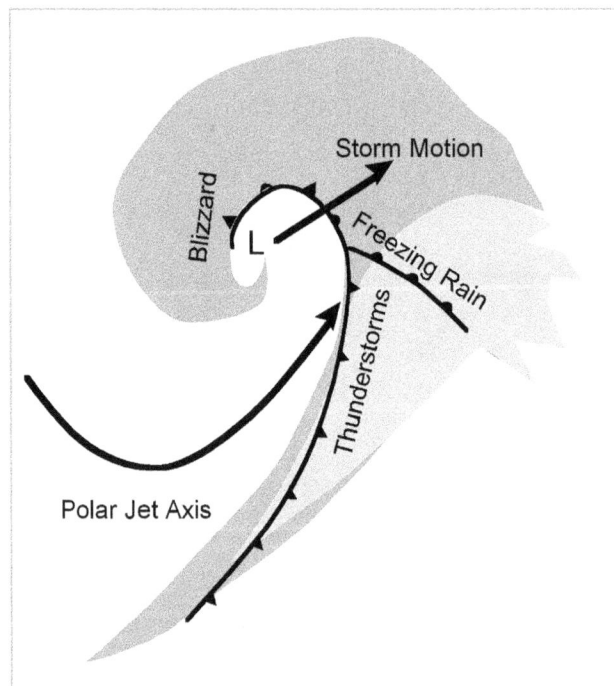

Figure 7.10 Schematic diagram showing the relationship between cloud shields, fronts, and weather hazards associated with a mature winter storm.

ACTIVITY

OBJECTIVE: To introduce the art of weather satellite image interpretation applied to a hazardous winter storm.

PROCEDURE: Use the schematic in Fig. 7.10 and the images in Fig. 7.11 as a guide to completing this exercise on Fig. 7.11a.

a. Analyze the fronts associated with this cloud pattern. Use blue for the cold front, red for the warm front and purple for the occluded front.

b. Indicate the location of the surface low-pressure center.

c. Indicate the location of the axis of jet stream aloft.

d. Starting at the low center position, indicate with an arrow the likely direction of motion of the storm.

e. Label the three classes of weather hazard commonly associated with winter storms, severe thunderstorms, freezing rain/sleet, and blizzard conditions.

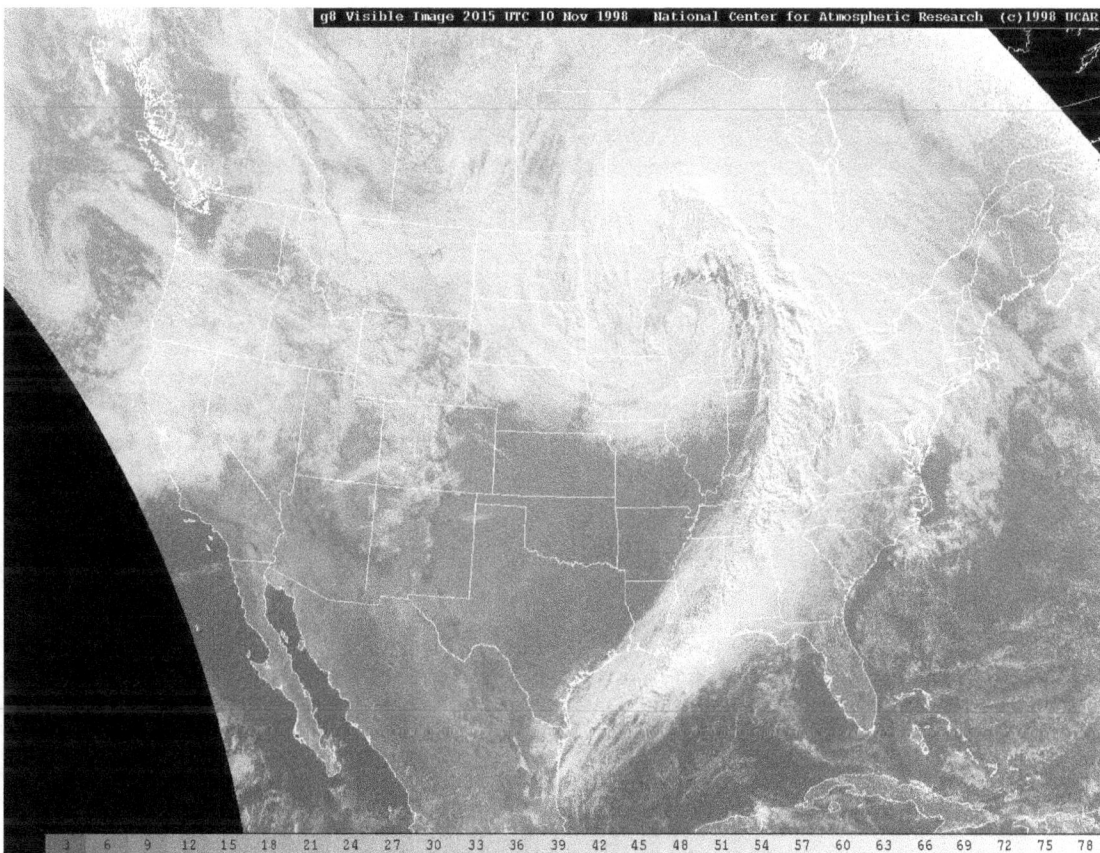

Figure 7.11a Visible satellite image at 2015 UTC 10 November 1998.

Fig. 7.11b Enhanced infrared satellite image at 2015 UTC 10 November 1998.

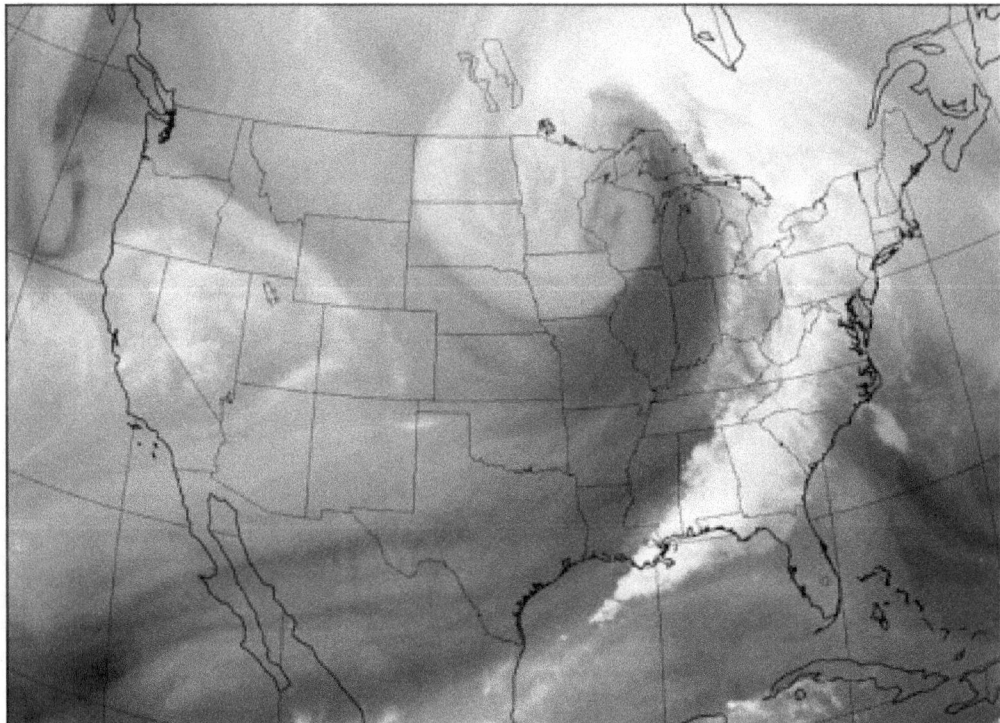

Fig. 7.11c Water vapor satellite image at 2015 UTC 10 November 1998.

QUESTIONS:

1.   What clues does the visible satellite image provide for the presence of thunderstorms?

2.   What clues does the infrared satellite image provide for the presence of thunderstorms?

3. Can you find the axis of the polar jet stream and upper-level low center in the water vapor image?

Lab 36: Numerical Weather Prediction

INTRODUCTION

When meteorologists were probing the atmosphere to understand the structure of storms at the turn of the last century, it was somewhat analogous to a blind man tapping his cane on an elephant to understand its nature. The ability to observe and describe the structure of something, whether it is an elephant or a storm system depends on the distribution of available data points.

The minimal resolvable wavelength is equal to two times the spacing of available data points (Fig. 7.12). Thus, the density of the observational and model grid needs to increase in accordance with the scale of the phenomena that is being simulated. To model a planetary wave aloft, with a wavelength of 10,000 km, grid spacing of 500 km may provide good insight. To model a thunderstorm, grid spacing of 0.5 km or less may be needed.

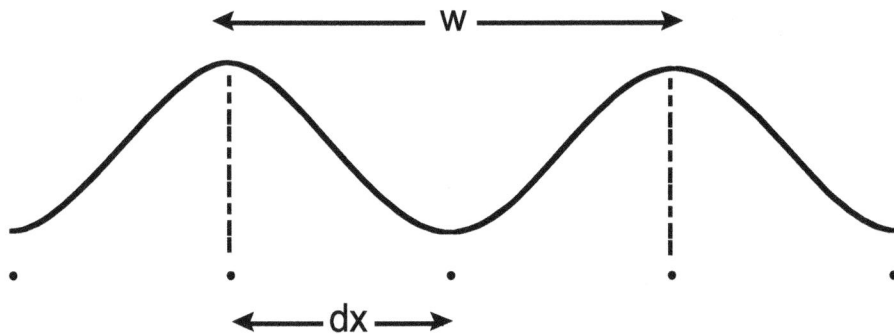

Figure 7.12 Schematic illustration of the minimal resolvable wave for a given data spacing.

We cannot measure everything, everywhere, all the time because of cost considerations and our inability to process the resulting data stream. The nature of our observations determines to a large extent our ability to forecast the weather. Numerical Weather Prediction (NWP) models rely on observations made in the upper atmosphere by radiosonde instruments carried aloft by weather balloons. Figure 7.13

shows the distribution of radiosonde release sites in North America. Note the lack of data over the oceans. Methods have been sought to solve the atmospheric data collection issues over the oceans for over thirty years. A most successful solution for the north Pacific was Environment Canada's introduction of a weather ship, known as *Station PAPA*, during the 1970s. The purpose of Station PAPA was to be a fixed data source about one weather day west of British Columbia. As such, it took not only daily ocean surface weather data, but could release daily radiosondes to gather regular upper atmospheric data. The major drawback of Station PAPA was the very high cost of maintaining a permanent crew and ship at a fixed point in the ocean. In 1981, Environment Canada made the decision to terminate all weather ships due to their high cost. Since the termination of Ship PAPA, a suitable replacement for upper air observations has been the subject of active research and development programs in Canada and the USA.

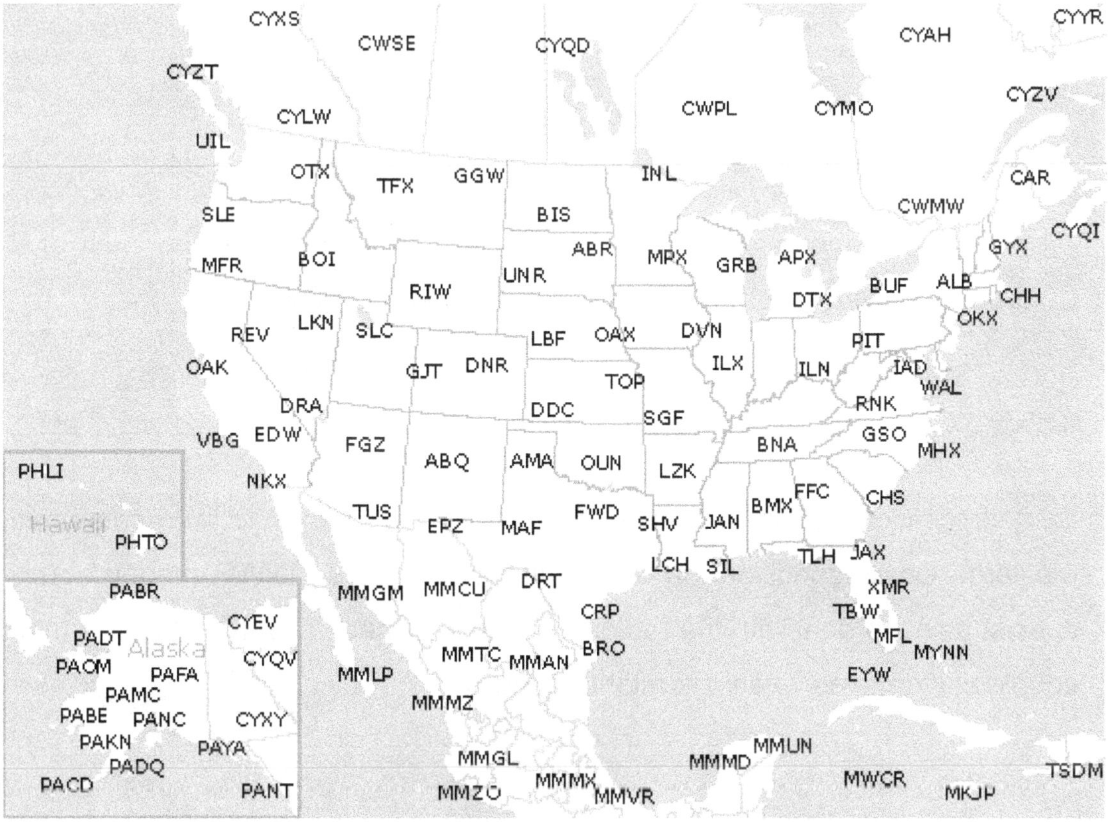

Figure 7.13 Map with locations of radiosonde sites labeled.

ACTIVITY

OBJECTIVE: To explore the relationship of observational data and grid spacing on NWP.

PROCEDURE:

1. On Fig. 7.13 sketch height contours for a trough and ridge pattern over the Midwest that corresponds to the minimal resolvable wavelength that can be observed in the upper troposphere with the current radiosonde network.

QUESTIONS

1. What is the wavelength in kilometers of the minimal resolvable weather system over the Midwest? Explain your answer.

2. Why do you suppose Canadian Ship PAPA was stationed to the west of British Columbia, rather than off the coast of Nova Scotia?

2. Sometimes local conditions cause the radiosonde data not to be representative of the larger-scale conditions in the upper atmosphere. NWP models must take special pains to deal with the radiosonde data in these events. Can you think of a couple of examples when this might be the case?

3. What observation system is most important for NWP in a remote ocean location like Hawaii?

4. Explain why three-day NWP forecasts of Washington D.C. should be more accurate than those for Seattle, Washington, even though both locations are near large oceans.

Lab 37: Weather Forecasting

INTRODUCTION

There is no better way to become familiar with the mechanics of forecasting the weather than by actually making a forecast of some aspect of the weather.

ACTIVITY

OBJECTIVE: To introduce the art and science of weather forecasting.

PROCEDURE:

For this exercise access NWP charts and MOS[5] via the web to forecast High and Low Temperatures, and Probability of Precipitation for tomorrow, and the following day. NWS observations taken at the chosen sites are used for verification of the temperatures and precipitation forecasts[6].

The particulars of this lab can easily be altered at the discretion of the instructor. To maximize the relevance of this exercise, choose the NWS reporting station nearest the location of your school as the first site to forecast for. The second site can be any place of interest to the instructor and students for which weather data are conveniently available. The second site can, for example, be chosen on the basis of interesting weather changes that are anticipated in the region.

Forecast duration and deadlines can be set by the instructor, for example at the end of the lab period or the afternoon. For more extended forecast contests, it may be convenient to have students email forecast numbers and set up an automated verification procedure.

---

[5] Go to http://www.nws.noaa.gov/tdl/synop/products/bullform.all.htm for real-time MOS.
[6] Go to http://www.nws.noaa.gov/ and navigate to the nearest Forecast Office for verification data for your site.

214

QUESTIONS

1. Discuss the significant weather systems affecting each city, and provide a summary of the main forecast issues for each city.

2. Explain which products you found most useful in making your forecasts.

Forecast Form

Tomorrow's High Temperature for HNL	_____ °F
Tomorrow's Low Temperature for HNL	_____ °F
Tomorrow's Probability of Precipitation for HNL	_____ %
Tomorrow's High Temperature for SEA	_____ °F
Tomorrow's Low Temperature for SEA	_____ °F
Tomorrow's Probability of Precipitation for SEA	_____ %
Day-after Tomorrow's High Temperature for HNL	_____ °F
Day-after Tomorrow's Low Temperature for HNL	_____ °F
Day-after Tomorrow's Probability of Precipitation for HNL	_____ %
Day-after Tomorrow's High Temperature for SEA	_____ °F
Day-after Tomorrow's Low Temperature for SEA	_____ °F
Day-after Tomorrow's Probability of Precipitation for SEA	_____ %

Error Points

Temperature error points will be calculated from the difference between the actual recorded temperature and the forecast temperature using the following formula: Temperature Error Points = $(T_{verification} - T_{forecast})^2$ up to 25 points, and 10 points for every degree off from the verified temperatures thereafter. e.g., 1,4,9,16,25,35,45...etc.

Total precipitation amount for the 24-hour period beginning at 12Z tomorrow and 12Z the following day. The forecast will be made in percent chance of precipitation (0-100%). Verification is as follows:

NONE	=	0
TRACE	=	5
> 0.01"	=	10

Error points will be calculated by taking the square of the difference between the verification and your forecast of precipitation chance/10.

For example, if you forecast a 30% chance of precipitation (= 3) and the verified total at HNL for the 24 hour period was .02" (= 10), the error points would be: *Error Points* = (verification - forecast)^2 = $(10 - 3)^2$ = 49.

A lower score is better. Therefore, the goal is to minimize the number of points generated by your forecasts. There are two parts to this exercise, the numerical forecast and the written description. Greater weight is generally given to the written part to encourage thoughtful forecasts and avoid unduly penalizing inexperience.

Chapter 8 Severe Thunderstorms and Hurricanes

'If the phone doesn't ring, you'll know that it's me. I'll be out in the eye of the storm.'
Jimmy Buffett

Thunderstorms occur when the atmosphere is unstable and the air is sufficiently humid to provide a fruitful source of latent heat. In some circumstances, thunderstorms can continue growing until they penetrate into the stratosphere, where stability through warmer temperatures aloft inhibits further development. There still is debate about the mechanisms by which a cumulonimbus cloud becomes electrically charged. Most experts believe that the charging results from the interaction of ice particles and *supercooled* water drops within the updraft region of the cloud. The end result is a cloud containing mostly positive electric charge in its upper part and mostly negative charge in its lower part. When the accumulation of electric charges becomes sufficiently large to overcome the insulating properties of air, a huge electrical spark or arc occurs and is called lightning.

Thunderstorms are most frequent during the warm seasons of the year and, for the most part, are beneficial to most environments. Adequate thunderstorm rainfall during the growing season is crucial over the grain belts of the world. Unfortunately, when thunderstorms are unusually severe, the attendant winds and precipitation can cause widespread damage to most types of vegetation.

Severe Local Storms

The term severe storm is commonly used in reference to any weather system that presents an immediate threat to life or property. Large, violent thunderstorms can produce a wide variety of weather phenomena that might be considered *severe*, including flash floods, lightning, hail, and strong winds, and on some occasions produce tornadoes. Other weather-related phenomena that may be considered a hazard to life and property include storm surges, freezing rain, and blizzards. However, to meet the NWS criteria for a *severe thunderstorm* or *severe local storm* the storm must produce at least one of the following, (i) a tornado, (ii) damaging winds or measured gusts of greater than 50 knots (55 mph), or (iii) hail with a diameter > 3/4 in. (1.9 cm).

The NWS criteria are emphasize here, because it is beneficial to have an understanding of the nature of the hazard when the NWS issues watches or warnings for severe thunderstorms.

Deep Convection

Almost all severe thunderstorms are associated with deep convection. To achieve deep convection there are three necessary ingredients. (i) A moist layer of sufficient depth in the lower or middle troposphere. (ii) A steep enough lapse rate above the moist layer to allow for a substantial "positive area" (or CAPE) between a parcel temperature and the environmental temperature. (iii) Sufficient lifting of a parcel from the moist layer to allow it to reach its level of free convection (LFC).

Once deep convection is found to be possible, the issue of the potential for severe events to occur with this convection becomes primary. To assess the potential for deep convection, the forecaster must be able to analyze the current stability of the troposphere and forecast changes resulting from thermal and moisture advection and vertical motion fields. In addition to the twice-daily radiosonde reports, the use of numerical weather prediction products from the National Centers for Environmental Prediction (NCEP) may be used to define areas in which deep convective development is possible.

A necessary ingredient for the development of large hail is a strong updraft, one that is capable of supporting the weight of a hailstone long enough for it to reach a large size. Ample instability (large positive area or CAPE) is a primary contributor to strong updrafts. In general, the greater the thermal buoyancy of the environment is, the greater the potential for large hail. The maximum hailstone size attained aloft can be linked at least partially to the strength of the updraft.

Damaging straight-line winds associated with deep convection almost always are generated by outflow that occurs at the base of the thunderstorm downdraft, also known as a *downburst*. Ingredients necessary for damaging winds at the surface are ones that promote very strong downdrafts, which include heavy precipitation loading and evaporative cooling associated with dry air at mid levels. Downward transport of momentum from the strong winds aloft that are entrained in the downdraft also enhance the outflow.

Tornadoes

Tornadoes, probably the most feared of violent weather events, are usually associated with organized thunderstorms called *supercells*. Supercells are characterized by a highly organized rotating updraft, accompanied by a strong downdraft. Supercells are likely when the environmental wind shear is strong (> 50 kt over 0-6 km above ground level) and the buoyancy is large (CAPE > 2000). Recent research in the U.S. suggests that only 20-30% of supercells produce tornadoes. However, the same study found that such storms almost always produce severe weather in the form of hail or high winds. Especially long-lived supercells (> 4 hrs) develop in environments with deeper, stronger wind shear, and these storms tend to be more isolated than the shorter-lived supercells. Tornadoes appear as grey funnels, cylinders or ropes extending from a cloud base to the ground. Most often they are small, less than a few hundred meters in diameter, and are visible for only a few minutes. The most violent tornadoes, representing fewer than 10% of the storms (about 50 per year in the United States), do most of the harm.

Tornadoes occur in many countries, but not with the frequencies or intensities experienced in the United States. On 3 May 1999 near Bridge Creek, OK, a research radar remotely sensed tornado wind speeds above ground level as high as 318 mph. Tornadoes are most frequent in spring and early summer and are common over a broad area extending from Texas to Michigan. This area is known as "Tornado Alley". The frequency of severe thunderstorms over the United States is related to geography: the north south orientation of the Rocky Mountains funnels Canadian arctic air southward toward the warm Gulf of Mexico, causing air masses with very contrasting characteristics to interact.

Many violent tornado outbreaks occur during "classic" synoptic conditions that are conducive to supercell development (Fig. 8.1). The favorable ingredients for supercells in these environments are lots of instability and *vertical wind shear*. Vertical wind shear is defined as a change in wind direction and speed with height. The classic tornado out break pattern is characterized by i) a strong extratropical cyclone or winter storm, ii) a diffluent upper-level trough to provide lift, iii) enhanced wind shear in the low to mid levels, iv) a mid-level intrusion of dry air (helps produce a strong downdraft, and v) sufficient instability. The combination of these factors together result in a "Big tornado day."

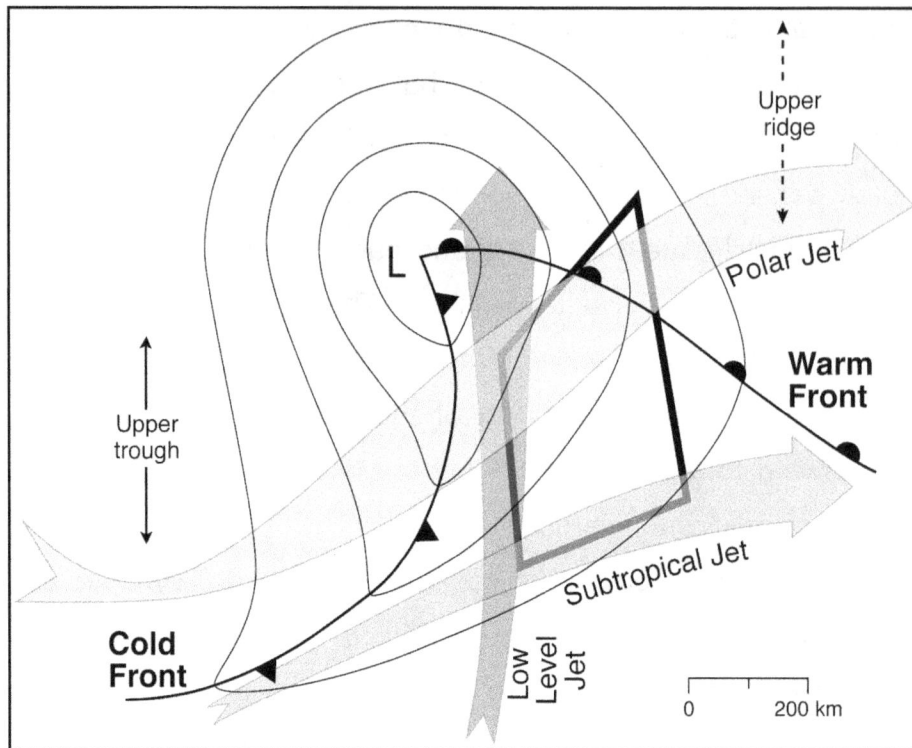

Figure 8.1 An increase in width along the shaded arrows indicates rising motion, a decrease indicates sinking motion. The bold box outlines the area prone to severe thunderstorms.

Tornado forecasting is really a matter of *nowcasting*, or very short term forecasting, because tornados form so quickly. Data from weather balloons (radiosondes), radars, and satellites are the key tools for nowcasting. Radiosonde and aircraft soundings can reveal large instability and wind shear that are characteristic of supercell environments. Radars can see the impact of rotation on reflectivity, resulting in the characteristic hook echo. Doppler radars can see wind shear over short distances, characteristic of the rotating updraft of a supercell. Radars can track storm motion and reveal cells that do not conform to the motion of surrounding cells. Satellite imagery can also identify unusual storm motion and document rapid anvil growth, indicative of strong updrafts.

The National Weather Service issues a *tornado watch* when a funnel is expected to form over a given region during the next few hours. When a tornado is sighted visually or by means of radar a *tornado warning* is issued. Communities in tornado-prone areas should adopt safety preparedness plans and conduct drills on how to react when a tornado watch or warning exists.

Hurricanes

Hurricanes form over warm, tropical oceans and are the most destructive storms on Earth. In 10 minutes a hurricane releases more energy than the entire world's nuclear weapons combined. In the western North Pacific, the storms that annually threaten the Philippines and Japan are called typhoons. In India and Australia, the same type of storm is called a cyclone. Tropical cyclones with hurricane-speed winds develop over the Indian Ocean and sweep into countries with low lying coastal zones such as such as Bangladesh and India, with the results of extensive flooding and numerous fatalities. Despite their destructive potential, hurricanes also produce a great deal of beneficial rain. Tropical cyclones are not classified as hurricanes until their maximum wind speeds exceed 65 kt (74 mph). In very strong hurricanes, the winds can exceed 200 kt (224 mph).

A hurricane is characterized by a small central region known as the eye, within which the winds are light and there are few clouds (Fig. 8.2). The eye is usually 20 to 40 km in diameter. Winds increase rapidly as one moves out of the eye and into a surrounding ring of thunderstorms referred to as the *eye wall*; maximum speeds are generally found in the eye wall at a distance of ~30 km from the center of the storm. A hurricane's overall diameter is typically between 500 and 1000 km.

There are five prerequisites for hurricane formation:
1. Warm ocean water with a temperature > 80° F (26° C) to a depth of ~50 m (deep thermocline), so that cooler water cannot easily be mixed to the surface by winds.
2. A pre-existing disturbance with cyclonic circulation (large low-level vorticity) persisting >24 hrs. As the air in the disturbance converges, angular momentum is conserved and the wind speed increases.
3. Small wind shear or little change in the wind speed or direction with height in the vicinity of the developing storm (wind shear < 20 kt from 850-200 mb), keeping warm, moist air over the storm.

223

4. Unstable troposphere characterized by enhanced thunderstorm activity. CAPE>1000 (Final CAPE in the eyewall is rather modest.)

5. Large relative humidity in the middle troposphere (no strong downdrafts). Moist air weighs less than dry air, contributing to lower surface pressures.

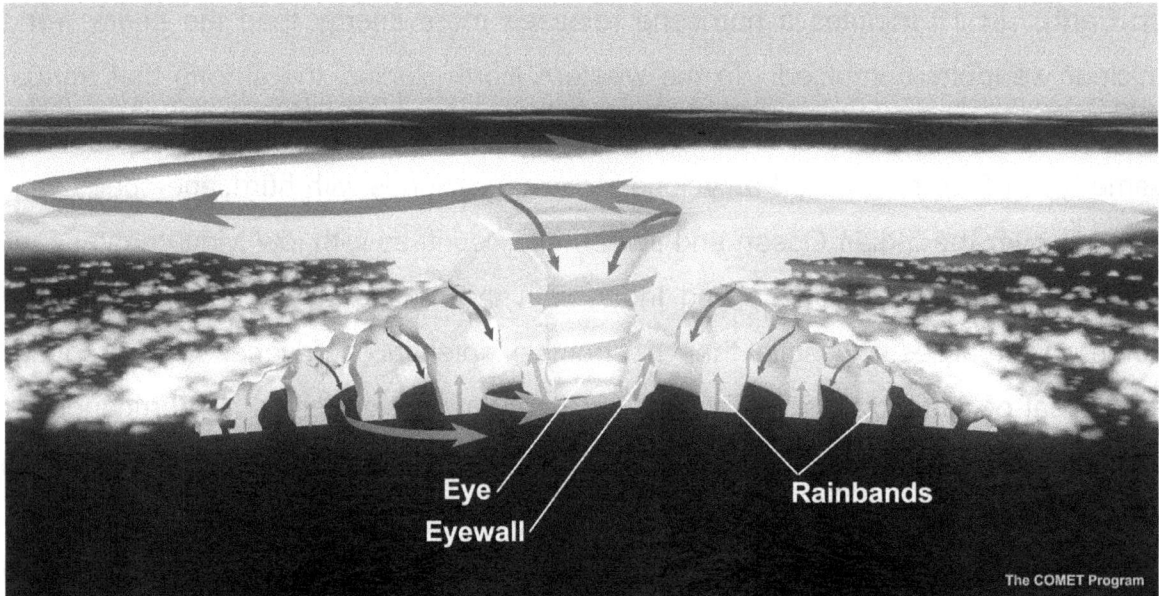

Figure 8.2 Schematic of hurricane showing wind flow and cloud structure (adapted from COMET Program).

The energy source for hurricanes is the release of latent heat in the storm clouds. Given a source of moisture from a warm ocean surface, hurricanes typically last for a week to 10 days. The ones affecting the eastern United States grow in intensity over the warm water of the Caribbean Sea and the Gulf of Mexico. The storms commonly are carried westward in the trade winds, but, as they approach the United States, they tend to curve toward the north and then northeast in part due to the influence of the prevailing westerlies at higher latitudes. When a hurricane moves over land or cold water, the supply of energy is reduced, and the wind speeds diminish. Over land, frictional forces also act to weaken the storm.

When a hurricane vortex moves over the ocean, the winds create large waves. Along coastlines, the winds cause an increase in the water level and flooding of low-lying coastal lands. Such a wind-induced, abnormal rise of the sea, called a *storm surge*, is historically responsible for most of the hurricane fatalities and damage. Heavy rains and strong winds supplement the effects of the storm surge. When hurricanes move over mountainous regions, the orographic lifting can lead to torrential rains and flooding, even at inland locations.

Weather satellites can effectively detect and track hurricanes over the entire Earth. When the storms get within a few hundred kilometers of land, radar and specially instrumented airplanes can also be used to observe a storm's intensity and the path it is following. When a hurricane is approaching a coastal location, early evacuation to higher ground of those people susceptible to the destructive force of the storm surge is essential. The National Weather Service issues a hurricane watch when there is a possibility of landfall within 36 hours. A *hurricane warning* is issued when landfall is likely within 12 to 24 hours. Large cities along the Gulf of Mexico are especially vulnerable to hurricanes and should be adequately prepared for them.

Lab 38: Investigating Hail

INTRODUCTION

Hail particles are ice pellets that are greater than five millimeters in diameter. As mentioned previously, hail forms in thunderstorm cells where there are strong updrafts, great vertical development. It is an interesting trick of nature that allows hail to form. That trick is the fact that cloud droplets in the atmosphere usually remain liquid even at temperatures well below freezing. In the absence of freezing nuclei to initiate crystallization, a clean liquid water droplet surrounded by air will not freeze until it is cooled to around −40° C! Liquid droplets at temperatures below freezing are called *supercooled*. There are special particles in the air that mimic the shape of ice crystals and fool the water drops into freezing. These are called *freezing nuclei.* Of course ice crystals make the best freezing nuclei. In hailstorms, strong updrafts exist that can catch ice particles and carry them back upward into the cloud. Many additional supercooled cloud droplets crystallize on the ice particle during this journey, adding a new layer of ice each time the hailstone rises and falls again. This cycle is responsible for the layered growth observed of hailstones. The cycle continues and the hailstone grows in size until the force of gravity overcomes the force of the updraft.

ACTIVITY

OBJECTIVE: The purpose of this experiment is to investigate an interesting twist of nature that is essential to the production of hail. The same principles demonstrated here also apply to the formation of most rain over the United States, in that 95% of our rainfall begins in the upper parts of clouds as snow. As snow crystals fall they sweep up supercooled droplets in a process called *accretion.*

MATERIALS:
§ crushed ice
§ large test tube
§ 400-600 ml beaker

§ salt

§ distilled water

§ thermometer

§ stirring rod

PROCEDURE:

1. Wash the test tube, being sure that no dust or dirt remains on the inside.

2. Fill the beaker not quite half full of water and add the salt. Pour in enough salt so that after stirring, you can still see salt on the bottom of the beaker. It is better to have too much salt than not enough.

3. Now add crushed ice to the mixture so that the beaker is nearly full; leaving room for the test tube.

4. Fill the test tube with cold distilled water so that the level of water in the test tube is just below the level of water in the beaker when the test tube is placed in the beaker (Fig. 8.3).

5. Put the thermometer in the beaker and then put the test tube in the beaker (Fig. 8.3).

6. Stir occasionally with the stirring rod for ~6 minutes. The temperature in the beaker should fall below zero. If the temperature does not fall below 0° C (32° F) you can add more salt and ice.

7. Once the temperature in the mixture has clearly fallen below freezing, record the temperature in the Data Table.

8. Remove the test tube and immediately drop a small piece of crushed ice into it. Record your observations in the Data Table.

8. Empty the test tube and repeat steps 3 through 7.

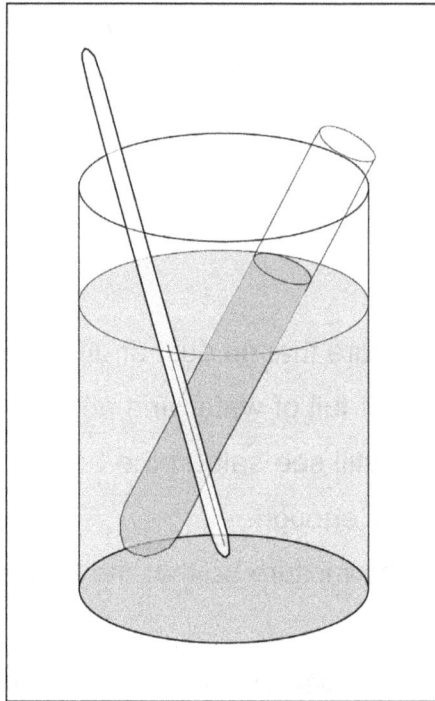

Figure 8.3 Schematic diagram

Data Table

Trial	Temperature (°C)	Observations
1		
2		

QUESTIONS:

1. What was the temperature in the beaker at the end of six minutes? What physical principle caused the temperature to drop below zero?

2. What happened when the piece of ice was placed in the test tube? Explain.

3. The water in the test tube was below its freezing point before the piece of ice was inserted. Why do you think that the water did not freeze before the ice was inserted?

4. Why do you think it is important to clean the test tube so well before you use it for this activity? Why did you use distilled water?

5. In this experiment, "hail" formed in the test tube. How does this compare with what happens in the atmosphere?

Lab 39: Voices in the Heavens

INTRODUCTION

The cause of thunder is one of the oldest riddles of recorded scientific speculation. Three centuries BC Aristotle published the first thunder theory. He believed that thunder was the result of air smashing against the clouds and, as the air struggled its way through the clouds, it kindled a flame that was lightning. In the mid-seventeenth century, British physicist Robert Hooke deduced that the duration of thunder was dependent upon the distance between the lightning stroke and the observer. Benjamin Franklin, who proved in his famous kite experiment that lightning was an electrical discharge, reasoned that if a spark produced in the laboratory produced a loud snap, then lightning should also produce a sound. "How loud must be the crack of 10,000 acres of electrified cloud?" he wondered.

Many theories were proposed until at the beginning of the 20th century a consensus evolved which assumed thunder must begin with a shockwave in air due to the sudden thermal expansion of the ionized air (plasma) in the lightning channel. The only experimental support for this theory came from optical (spectroscopic) temperature determinations up to 36000 K, hotter than the surface of the sun. However, recent experiments with short atmospheric arcs of lightning strength revealed much higher temperatures than those found by lightning spectroscopy. Arc photography proved conclusively that the plasma did not expand thermally in all directions, but preferentially at right angles to the current, as if driven by organized electrodynamic action. Thus, despite being described as "well understood" in modern meteorology textbooks, the underlying cause of thunder remains a topic of speculation and active research.

Sound waves from all segments of the lightning stroke are produced almost simultaneously, typically over a time interval much less than a second in length. What variations we hear in a thunder peal result from the time required for the sound from different segments of the lightning bolt to reach our ears, the nearest segments being heard before the more distant. This time differential, coupled with the length and orientation of the larger segments of the lightning flash, determines the unique character of each thunder peal we hear.

The velocity of sound waves depends on the material through which they are traveling. As you might expect, sound waves travel through steel railroad tracks much faster than through water, and the speed of sound through water is greater than its speed through air. The greater elastic modulus of steel more than makes up for its greater density compared to water. Knowing the speed of sound through materials allows us to use sound waves to determine distances. The system used for finding distances in water is called sonar, short for sound navigation ranging. This system is used to find the depth of the ocean at various locations (map the ocean floor) and to locate objects underwater.

Sonar was invented in the 1920s by August Hayes. By bouncing sound waves off of objects, he found you could measure distance. The system is similar to those used by bats and porpoises to navigate. But before sonar can be used, the speed of sound waves must be known. The following experiment illustrates how the speed of sound waves can be found.

ACTIVITY

OBJECTIVE: Perform an experiment to determine the speed of sound (e.g., thunder) in air. Also, gain an understanding of how sonar works.

MATERIALS:
§ Two metal objects (1 kg weights work well)
§ A meter stick or tape measure
§ A watch or clock with a second hand

PROCEDURE:
1. Find a large flat wall either indoors or outdoors. A school hallway with no side corridors works indoors or a gymnasium wall does fine outdoors. You will be reflecting sound waves off the wall to determine the speed of sound (Fig. 8.4).

2. Position yourself 100 meters away from the wall. Measure the distance to make sure it is accurate.

3. Tap the two weights together and listen for the echo. You will need to be very quiet to hear.

4. Once you have produced a suitable echo, tap the weights together at a steady rate so that you hear the echo exactly halfway between taps. This will take a little practice.

5. Now that you have the rhythm of the experiment, you need to count how many taps you make in to seconds. Don't count the echoes. Repeat the experiment two more times. Record your results in a data table.

6. The speed of sound can be determined by using this equation: velocity = distance ÷ time. The distance that the sound wave traveled was twice the distance from you to the wall (round trip). The time it took was one half the time between taps.

7. Compute the speed of sound for the three trials and compare your results with a reference value for the speed of sound in air.

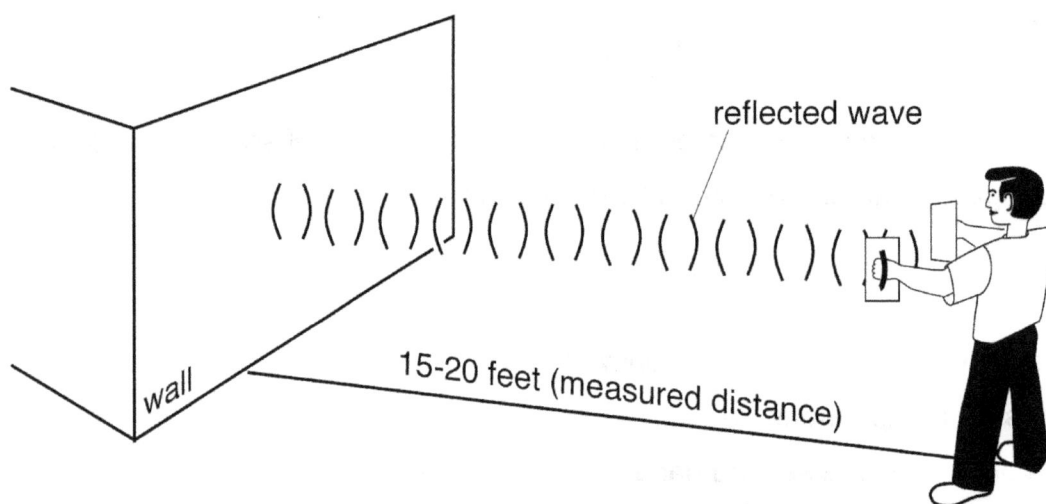

Figure 8.4 Schematic diagram showing method for finding the speed of sound in air.

QUESTIONS:

1. How far away would you estimate a bolt of lightning struck the earth if you heard the thunder eight seconds after you saw the lightning?

2. What is the minimum time it would take an autofocus camera using a sonic ranging device to determine the distance to an object that is 40 feet from the camera?

3. If a distant thunderstorm is 10 miles away, how long will it take for the thunder generated in the storm to reach your ears?

4. How much time would elapse between a sonar system's "ping" on the surface and the return of the echo if the depth of a lake is 450 meters?

5. How deep would the Atlantic Ocean be beneath your ship if the time between your sonar's ping and the return of the echo was consistently 4.5 seconds?

Lab 40: Storm Chasing

INTRODUCTION

Tornado chasers are in the hunt for supercells, the special breed of long-lived, rotating thunderstorms that tend to produce the greatest number and strongest tornadoes. All organized severe storms require some combination of moisture, instability, lift, and wind shear. A weakness in one may be compensated for by unusual strength in another; but with a glaring lack of any of those ingredients, storms may be weak or may not form at all. Surface and upper-air forecasts can be analyzed as far out as two or three days, to determine where these ingredients are setting up and where they may shift. Sufficient moisture, instability and lift for thunderstorms are rather common in spring and summer over tornado alley in the US Midwest, but to get supercells, there must be some *vertical wind shear*, a change in horizontal wind direction and speed with height. A change in speed (e.g., light surface winds and a jet stream overhead) will support severe storms, but a change in direction, particularly in the lowest 6,000 feet of the atmosphere, is an additional element that helps to sustain supercells. An excellent scenario for tornadic storms is where winds are blowing at 15 knots out of the east at the surface, out of the southwest at 40 knots at 5,000 feet, and out of the west at 50 knots at 10,000 feet above ground level.

Storm chasing as a hobby has seen considerable growth in recent years and the hazards imposed not only by storms, but increasingly by fellow chasers has become an issue. Safe storm chasing may seem like an oxymoron, but the fact is that a storm chaser's exposure to risk can be greatly reduced through a combination of knowledge and responsible behavior (see http://www.cimms.ou.edu/~doswell/Chasing2.html for an essay on storm-chasing courtesy and safety).

ACTIVITY

OBJECTIVE: A record number of tornadoes occurred across the United States during the first 10 days of May 2003. In this lab the goal is to use the weather charts and data provided to identify the best location to observe a tornado during the afternoon of

4 May 2003. Since this exercise is based on a past case, observed data and analyses are provided to diagnose the potential for tornadic thunderstorms. However, in practice chasers in the field would consult a range forecast products valid for the next day produced by NWP models and available on the web (e.g., http://www.rap.ucar.edu/weather/).

MATERIALS:
§ blank map of the US
§ analysis maps for several levels in the atmosphere valid for 12 UTC on 4 May 2003.

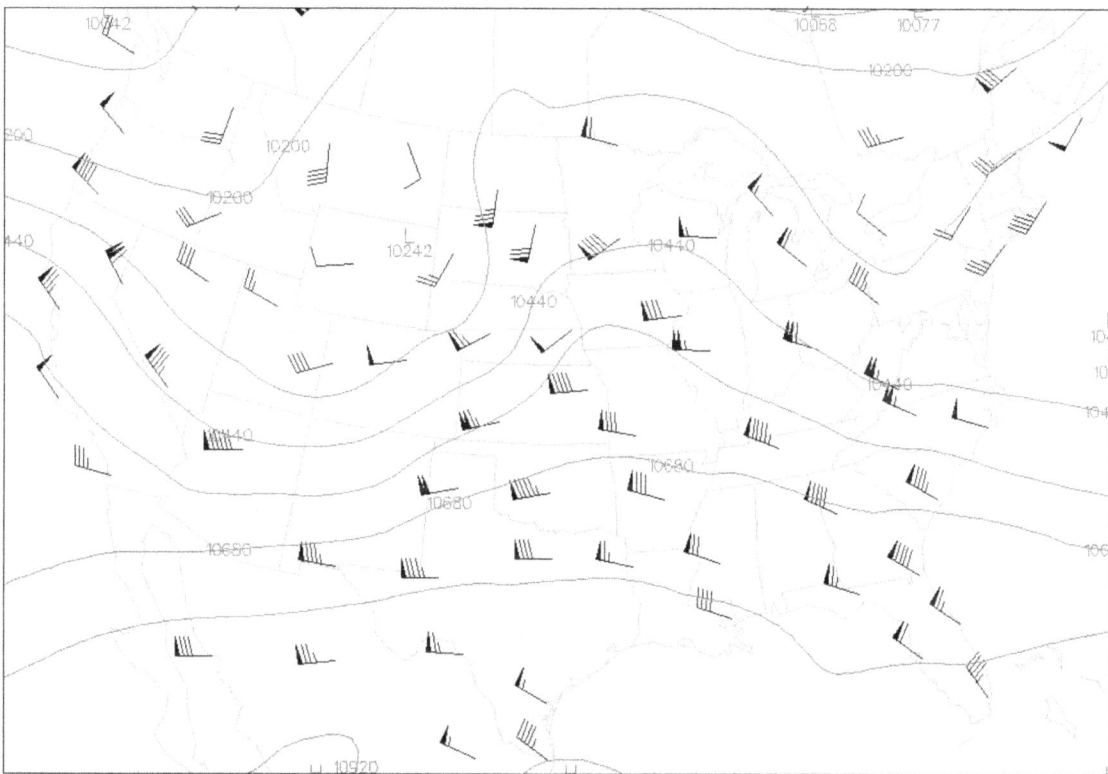

Figure 8.5 250-mb analysis for 12 UTC on 4 May 2003, showing heights (every 120 m) and wind barbs.

PROCEDURE

At this level in the upper troposphere, strong winds and divergence are analyzed. Jet streaks aloft are commonly associated with the occurrence of severe thunderstorms. Divergence aloft is associated with transverse circulations in the entrance and exit regions of enhance jet flows (see Lab 29 and Fig. 6.7).

1. On the above chart, outline in green the 110 and 90-kt isotachs associated with the jet streak that is over the southwestern USA.

2. Indicate the axis of the jet in this area with a heavy dashed blue line.

3. Draw a bold red circle with a diameter of ~5° latitude over the divergent left jet-exit region (see schematic of jet streak in Fig. 6.7). Find where the 110 and 90 isotachs cross the axis of the jet on the eastern side of the jet streak. The red circle should enclose the area to the north of the jet axis and include these two intersection points. Look for diffluence of the flow within circle.

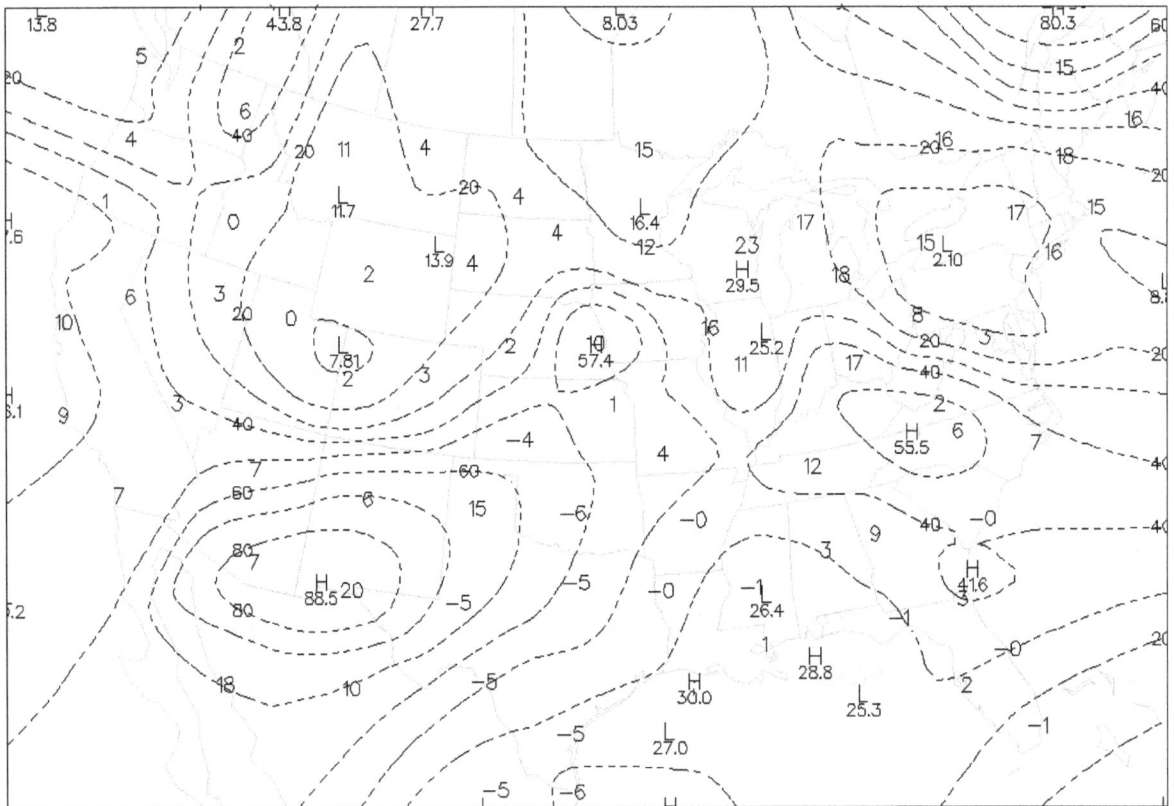

Figure 8.6 500-mb chart for 12 UTC on 4 May 2003, showing isotachs (every 10 kt) and plotted lifted index values.

PROCEDURE

At this level in the middle troposphere, the presence of high winds and the advection of cold air help create instability in the atmosphere (Fig. 8.6). One common way to diagnose instability is to look at values of the lifted index. The lifted index (LI) is the temperature difference between a parcel that is lifted to 500 mb (following dry and moist adiabats) from near the surface and that of the environment at 500 mb.

1. Outline the 50 and 60-kt isotachs in blue over the southwestern USA associated with the approaching jet streak. Divergence and vertical motion are associated with the right exit region of the jet streak.

2. Advection of cold air is important at this level for reducing the stability of the atmosphere. Draw contours for LI = 0 and LI = -3. Shade in red the region near the jet exit where the LI has values <-3, an indication that thunderstorms are likely.

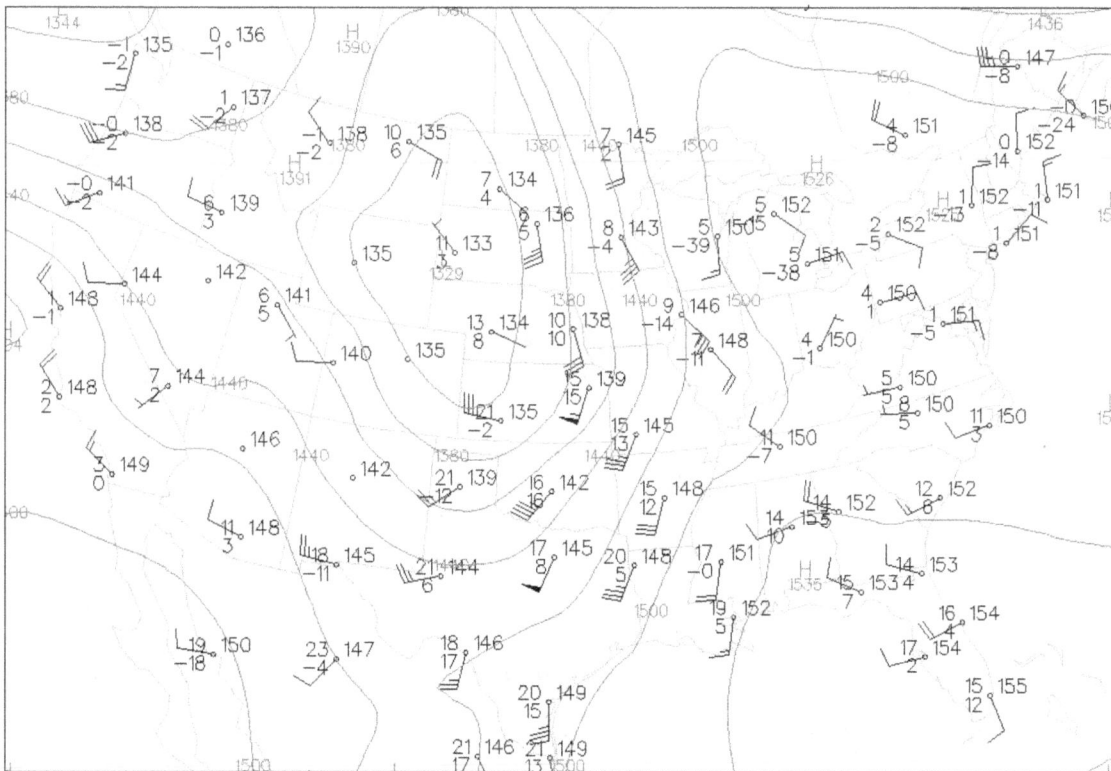

Figure 8.7 850-mb chart for 12 UTC on 4 May 2003, showing heights (every 60 m) and station plots.

PROCEDURE

At this level of the atmosphere evidence of moisture and the presence of a low level jet are analyzed (Fig. 8.7). The low level jet is critical for bringing warm moist air northward at low levels from the Gulf of Mexico.

1. Draw isotachs for wind speeds equal to 30 and 50 kt in blue.

2. To diagnose the presence of moisture at this level forecasters look at the dewpoint depression. The dewpoint depression is the difference between the temperature and dew point temperature plotted to the left of the station. ($D = T - T_d$). The smaller the dewpoint depression the closer the air is to saturation. Circle in red the region whose stations show a dewpoint depression of 2°C or less.

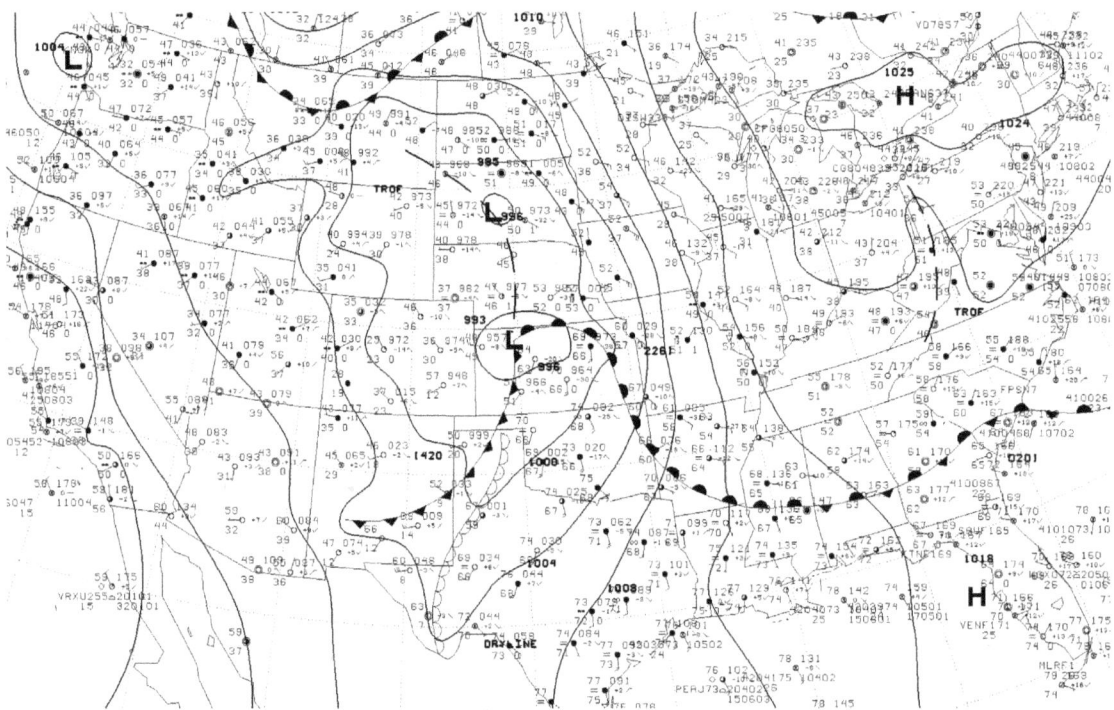

Figure 8.8 Sea-level pressure chart for 12 UTC on 4 May 2003, showing isobars (every 4 mb) and the location of frontal systems.

PROCEDURE

1. Outline the dry line in blue and shade the region east of the dry line with dew point temperatures > 64°C in red on Fig. 8.8.

2. On each of the previous charts, you were asked to mark features of particular significance to severe weather forecasting in red. On the map of the USA below (Fig. 8.9) overlay and label these special weather elements and note where they overlap. The overlays can be done directly on the blank map by eye, or the blank map can be copied to an overhead transparency for this purpose. Now superpose the axis of the upper level and low-level jets. Draw a box encompassing the region with the greatest number of overlapping features. Now book a motel in the center of the eastern side of that box and drive there before bedtime without breaking too many speed limits.

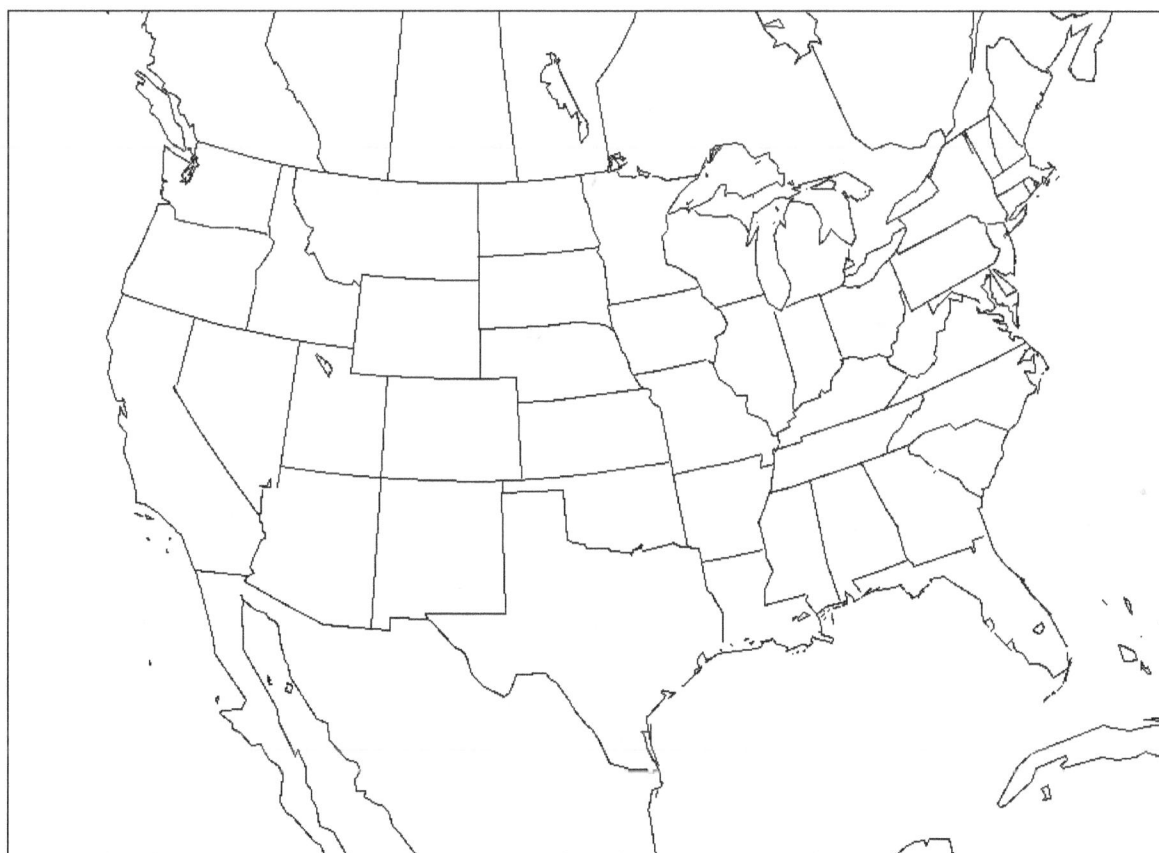

Figure 8.9 Blank map of the continental USA.

QUESTIONS

1. Why book a motel on the eastern side rather than the center of your box?

2. Look again at the 500-mb and surface charts (Figs. 8.6 and 8.8). Explain how the flow at these levels contributes to the low LI values over parts of the Midwest.

3. Based on your analysis in this exercise, discuss the reasoning you used to determine the most likely tornadic thunderstorm location for the following afternoon.

4. This lab relies on a limited set of products to make a forecast. What additional tools or products might a chaser consult in making plans for tomorrows chase?

Lab 41: Tracking Hurricane Iniki

INTRODUCTION

Because of their destructive nature, it is very important to accurately track and forecast the movement of hurricanes. The location of the center of a hurricane is obtained through a combination of observations from satellite, aircraft, ships, and islands. The motion thus obtained can then be extrapolated into the future and compared with computer simulations of the storm track in order to make predictions of the future path and speed of the storm. Watches and warnings are provided to emergency managers and the public by the National Weather Service based on the expected storm track to give people in affected areas time to prepare and evacuate as needed.

ACTIVITY

OBJECTIVE: The objective of this activity is to follow the track of hurricane Iniki for several days, with the goal of anticipating the storm's landfall. A second objective is to learn the meaning and implication of hurricane watches and warnings.

The National Weather Service issues a *hurricane watch* when there is a possibility of landfall within 36 hours. A *hurricane warning* is issued when landfall is likely within 12 to 24 hours. The destructive force of hurricane winds is proportional to the velocity squared. Meteorologists Saffir and Simpson codified the destructive potential of hurricanes with their scale provided in Table 1.

Table 1 Saffir-Simpson Hurricane Scale

Category	Max. Wind Speed (kt)	Max. Wind Speed (mph)
Tropical Depression	≤33	≤38
Tropical Storm	34-63	39-73
1	64-83	74-95
2	84-96	96-110
3	97-113	111-130
4	114-135	131-155
5	>135	>155

MATERIALS:

§ pencil

§ map

PROCEDURE:

In this exercise we will use nautical miles per hour (kt) for wind speed, instead of the usual mph, just as the National Weather Service does, because the distance in nautical miles traversed by a storm in one hour relates directly to the spacing of the latitude lines on the map that we will use to plot Iniki location data. Sixty nautical miles = 1° latitude.   Just remember that winds given in kt are about 10% stronger than the equivalent mph. So 10 kt ~11 mph.

1. Study the storm track data in the Table. It contains:
    a. date/time of the observation
    b. location of the hurricane's eye (latitude & longitude)
    c. maximum wind speed (*intensity*) of the hurricane (NOT the speed of storm motion)
    d. minimum sea-level pressure (MSLP), an alternate measure of *intensity*
    e. whether the pressure was *estimated* or measured in situ with an *instrument*

2. Plot the track data (LAT/LONG) in Table 2 on the map provided. For each point, write the minimum sea-level pressure. Mark the date for the beginning of each day (0000 UTC). Color code the track according to the Saffir-Simpson hurricane intensity scale.

3. Answer the accompanying questions.

Table 2 – Hurricane Iniki Track Data – September 6-12, 1992

Date/Time	Eye Position		Max. Wind	MSLP pressure	
(UTC time)	LAT	LONG	(kt)	(mb)	est./inst.
06/1800	12.2	-140.0	30	1008	est.
07/0000	12.3	-141.1	25	1008	est.
07/0600	12.3	-141.7	25	1006	est.
07/1200	12.2	-142.2	30	1006	est.
07/1800	12.1	-143.0	30	1004	est.
08/0000	12.0	-144.5	35	1002	est.
08/0600	12.0	-146.0	40	1000	est.
08/1200	12.1	-147.5	40	1000	est.
08/1800	12.3	-149.0	50	996	est.
09/0000	12.4	-150.2	60	996	est.
09/0600	12.7	-151.6	65	992	est.
09/1200	13.0	-152.9	65	992	est.
09/1800	13.4	-154.3	80	984	est.
10/0000	13.8	-155.5	85	980	est.
10/0600	14.3	-156.9	90	960	inst.
10/1200	14.7	-157.8	100	960	est.
10/1800	15.2	-158.6	100	951	inst.
11/0000	15.9	-159.3	110	948	inst.
11/0600	16.8	-159.8	115	939	inst.
11/1200	18.2	-160.2	120	938	inst.
11/1800	19.5	-160.0	125	938	inst.
12/0000	21.5	-159.8	115	945	inst.
12/0600	23.7	-159.4	100	959	inst.
12/1200	25.7	-159.0	80	980	est.
12/1800	28.1	-158.9	80	980	est.

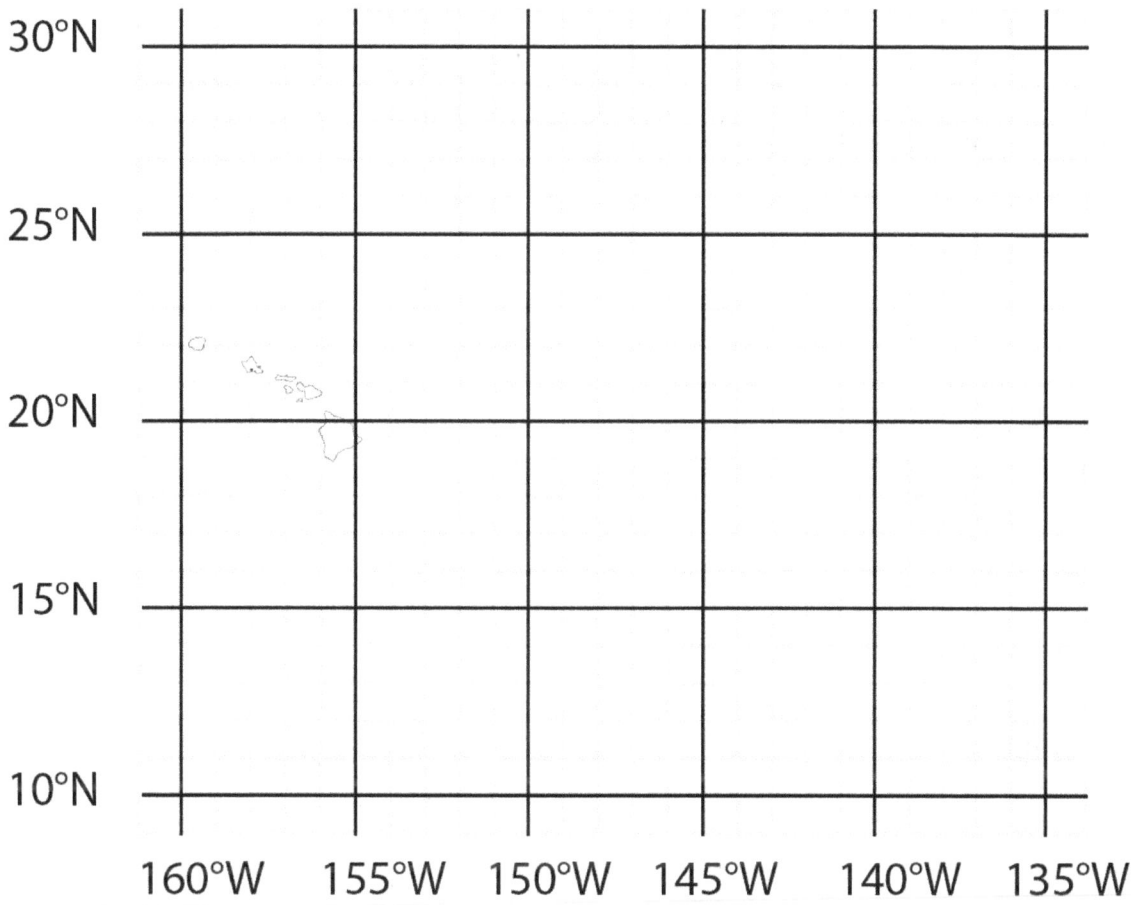

Figure 8.10 This map is a smaller version of the hurricane-tracking chart used at the National Hurricane Center.

QUESTIONS:

1. Based on the track you plotted, estimate the time (in HST) that Iniki's eye crosses the south coast of Kauai.

2. Based on the time you found in #1, when should the National Weather Service have issued its hurricane watch and warning?

a. Watch

b. Warning

3. How does the timing of the watch and warning fit logistically into most people's daily routines? Would you make any changes to account for this? Explain.

4. What was the hurricane's speed of motion (in kt) at the time that the **eye crosses Kauai**? What about when the **watch** and **warning** were issued? [remember: speed = distance/time]

5. Based on the speeds calculated in #4, what is the range of maximum wind speeds (in kt) as Iniki crosses Kauai? On which side of Kauai would you expect the greatest damage from high winds and storm surge?

6. Describe the relationship between MSLP and maximum wind speed over the course of Iniki's lifetime.

7. The strongest hurricanes (Category 5) have slower wind speeds than the stongest tornados (EF-5), but a hurricane will inflict more total damage – why?

Lab 42: Flash-Flood Forecasting

INTRODUCTION

The National Weather Service in Honolulu reported 322 flash flood events in a 40-yr period; an average of more than 8 flash floods per year. Flash floods cause the most direct weather-related fatalities in Hawaii, and have produced severe property damage. A flood in January 1980 on Maui resulted in >$50 million damage in one event. A flood on New Years Eve of 1987 caused > $35 million damage on Oahu. Flash floods are the result of heavy rains that fall on saturated soils. In Hawaii, and elsewhere, steep terrain and shallow soils contribute to the occurrence of flash flooding. Urbanization (increase in non-permeable surfaces) and blocked drainage channels can contribute to the hazard. An added danger in Hawaii and other locations is the presence of aging earthen dams.

Heavy rainfall is usually associated with deep convective clouds (thunderstorms) and slow storm motion. Deep convective clouds form in unstable environments with ample low and mid-level moisture and enhanced upper-level divergence. Terrain forcing aids in lifting the air mass and also in anchoring thunderstorms to prolong the downpours. In mountainous terrain a flash flood can occur within an hour of the start of heavy rainfall. Therefore, forecasting flash floods with much lead-time is a special challenge for forecasters. Radars, because they can observe the rainfall accumulation over an entire river basin at once, have proven particularly useful in monitoring the development of dangerous conditions. The NWS issues the following advisories for flash floods (i) Flood Potential Outlook (36 hr in advance) event is possible within 36 hr, (ii) Flood Watch (36 hr in advance) event is likely within 36 hr, and (iii) Flash Flood Warning (updated every 3 hr) threat to life and property is imminent or occurring.

248

ACTIVITY

OBJECTIVE: The goal of this activity is to understand the relationship between precipitation and topography in the formation of a flash flood event.

MATERIALS:

Map of O'ahu with plotted rainfall data (mm) for a heavy rain event that occurred 25 January 1996.

PROCEDURE: Draw contours of equal rainfall (isohyets) every 10 mm on Fig. 8.11.

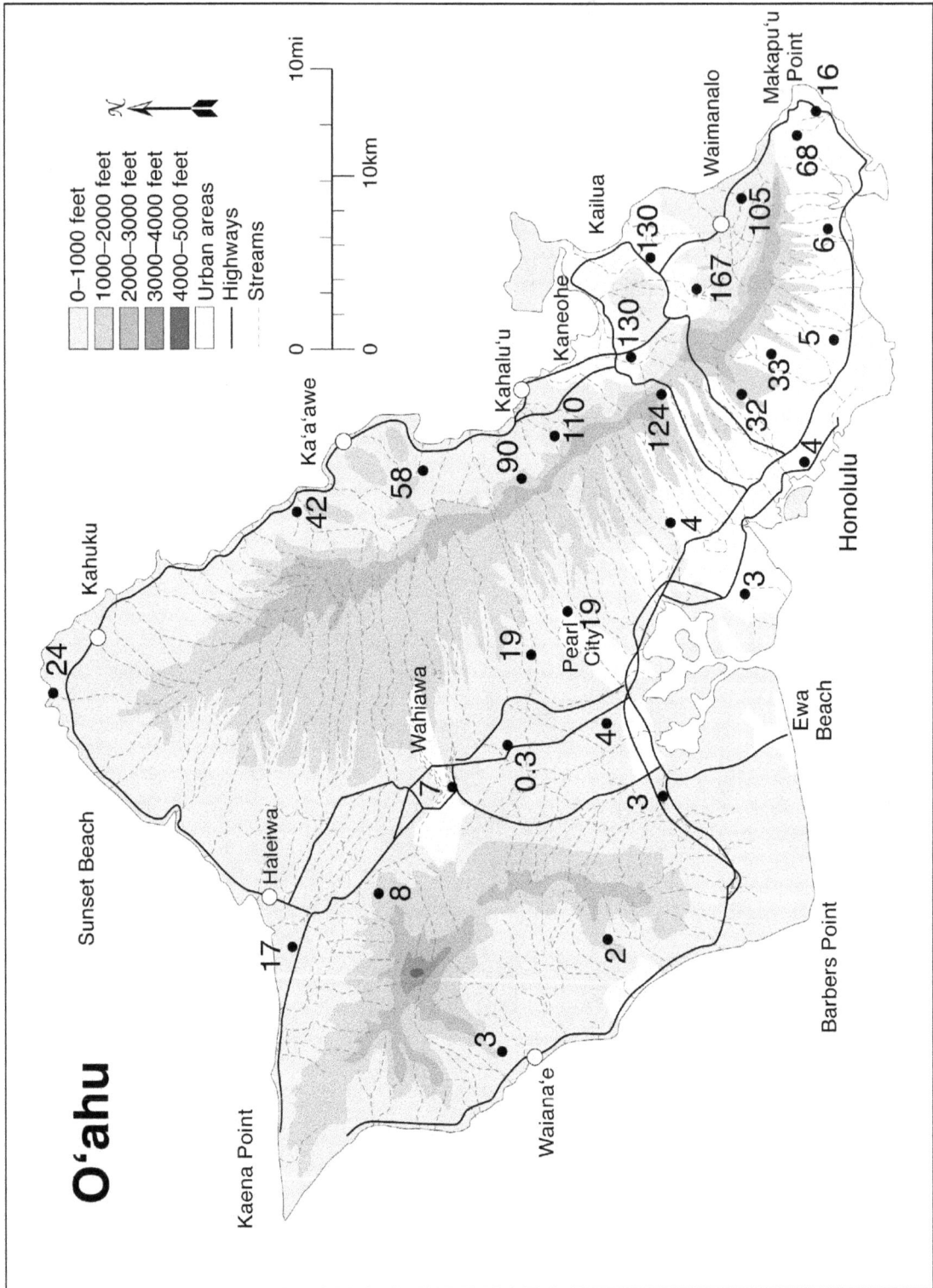

Figure 8.11 Map of O'ahu, HI with plotted 24-hr rainfall data for 25 January 1996.

QUESTIONS

1 Given the distribution of rainfall from this storm, what do you think the prevailing wind direction was during this event? Given that wind direction, in what direction was the center of surface low pressure located?

2. How did the shape of the mountains on O'ahu influence the rainfall distribution in this case?

3. Where on O'ahu was the threat for flash flooding greatest?

4. What impact might urbanization on O'ahu have had on the flooding in this case?

5. Having an urban population in an area prone to flash flooding increases the exposure to this weather hazard. Thinking about transportation, explain why the residents of O'ahu may be exposed to an extra hazard?

Lab 43: Hurricanes and Flooding

INTRODUCTION

Floodwaters resulting from heavy rains produced by tropical storms are the second most deadly hazard after the storm surge. Hurricane Mitch in late October of 1998 was a category 5 hurricane in the Caribbean Sea. Mitch hovered over Central America resulting in an extended period of torrential rains and mudslides. Mitch left ~10,000 dead, 7000 missing, and hundreds of thousand homeless. The flooding left up to 80% of infrastructure damaged and 70% of agricultural production lost. In the wake of the storm Honduras needed an new country map.

As introduced in Chapter 7, the strength of the signal scattered back to the radar by precipitation particles gives information on the rate of rainfall (Table 8.2) and this in turn provides important information on the structure of storm systems and the potential for hazardous flooding (Fig. 8.12).

ACTIVITY

OBJECTIVE: The objective of this activity is to use radar reflectivity data to investigate the structure of hurricane rainfall and flood potential.

PROCEDURE:

1. Use the information in Table 8.2 below to interpret the radar data for hurricane Hugo given in Fig. 8.12.

Table 8.2 Reflectivity Intensity

D/VIP Level	Echo Intensity	Estimated Precipitation	Rainfall Rate (inches/hour)
1	Weak	Light	< 0.2"
2	Moderate	Moderate	0.2 - 1"
3	Strong	Heavy	1 - 2"

2. Overlay tracing paper on the first panel of Fig. 8.12, mark the corners of the panel to register the paper, and shade areas with D/VIP levels =3.

3. Move the tracing paper to the remaining panels, register each, and repeat the shading. The result will be a composite map of the heaviest rainfall. Use your composite rainfall map answer the question 1, which follows.

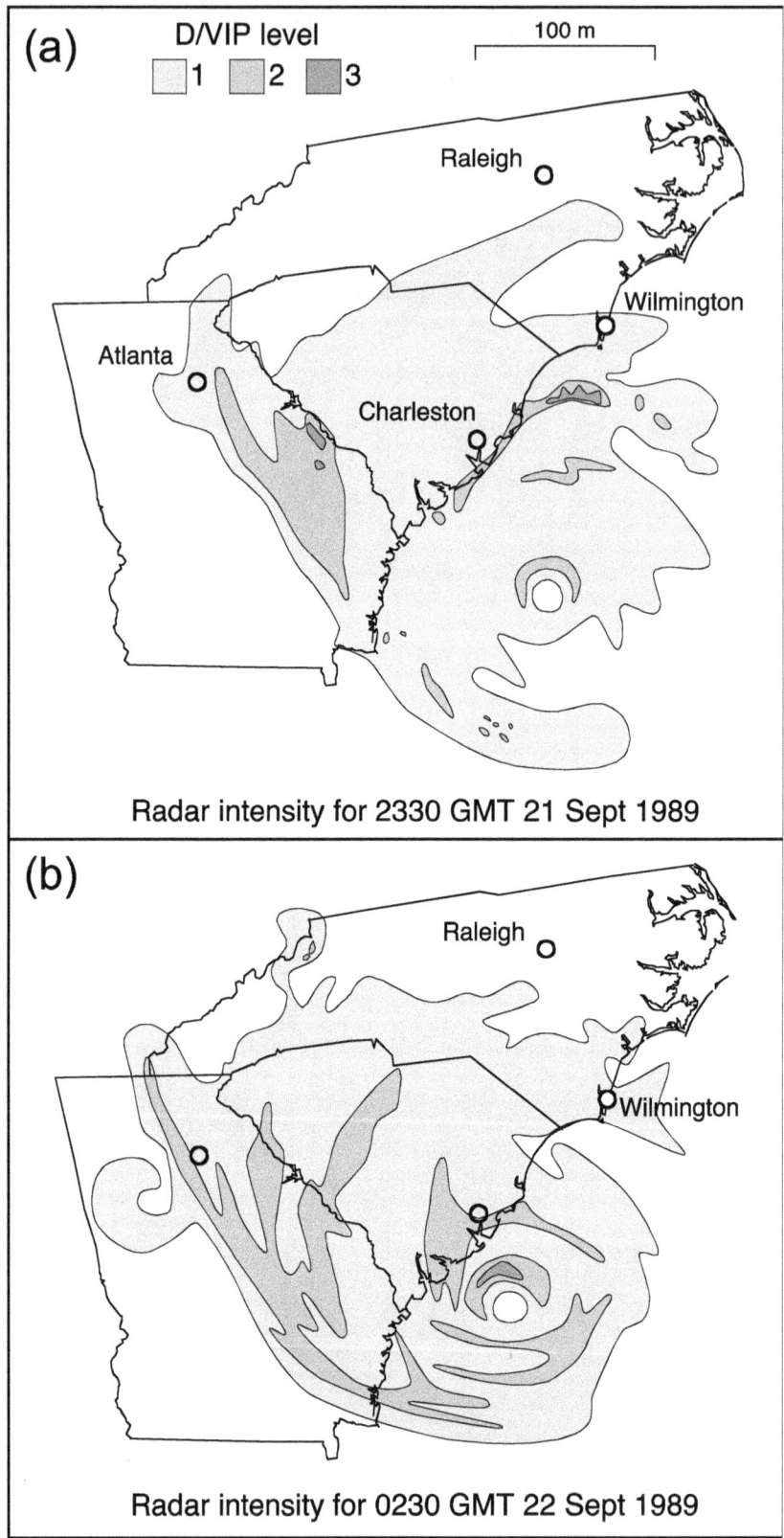

Figure 8.12 Radar intensity for (a) 2330 on 21 September 1989 and (b) 0230 on 22 September 1989.

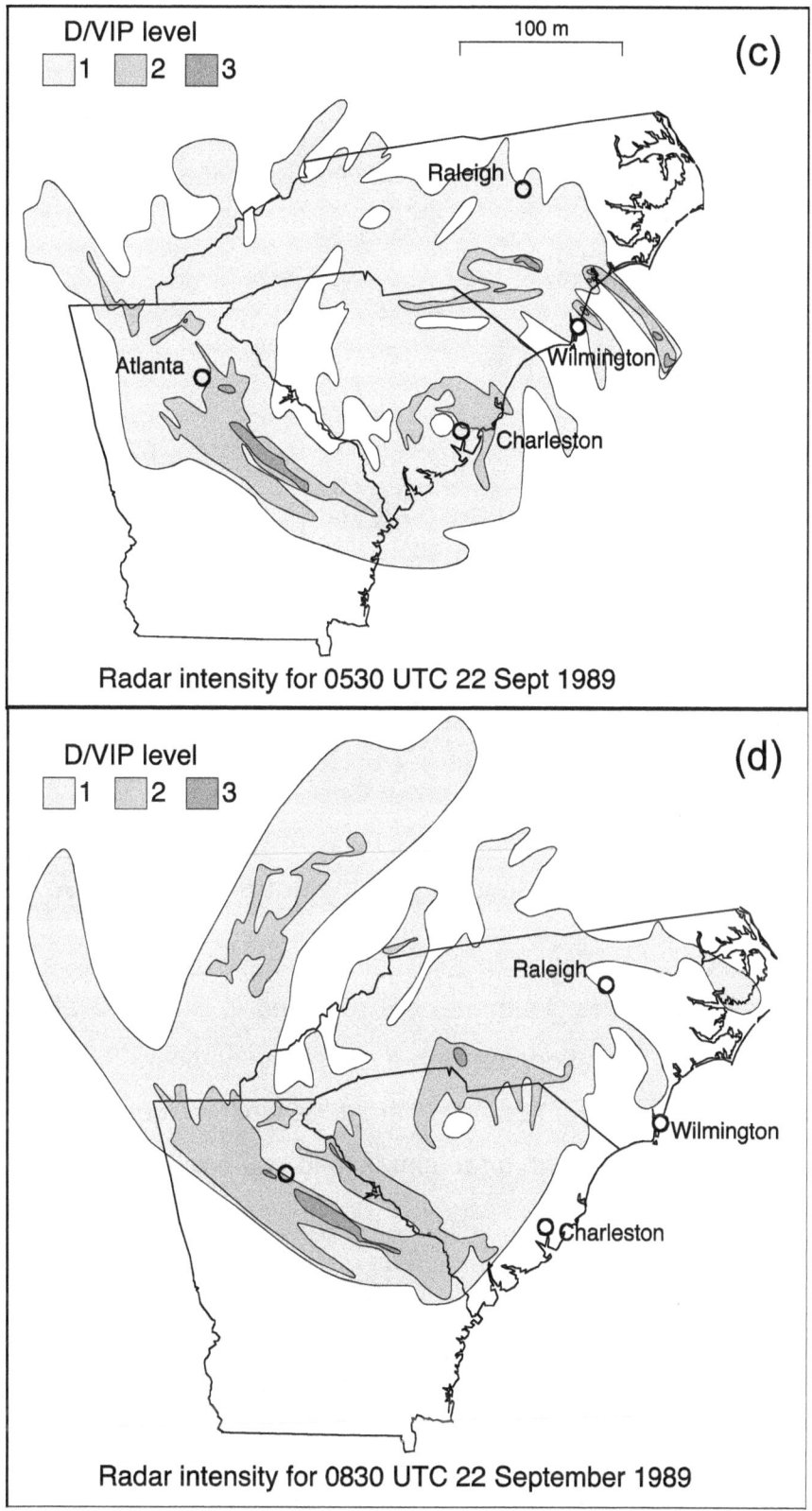

Figure 8.12 Radar intensity for (c) 0530 on 21 September 1989 and (d) 0830 on 22 September 1989.

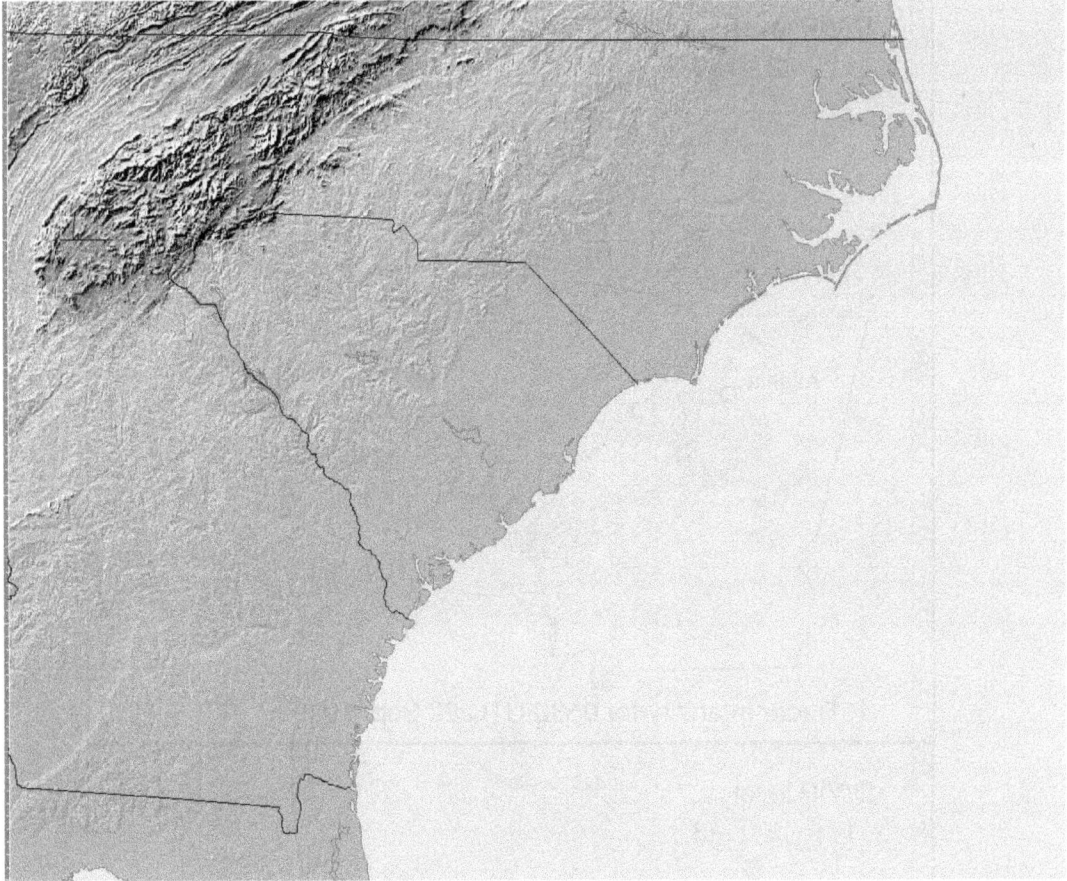

Figure 8.13 Relief map of the southeastern USA.

QUESTIONS:

1. On the relief map of the Southeast US provided in Fig. 8.13 above, indicate the area most prone to flash flooding with a red circle. Indicate the area most prone to an extended period of river flooding with a blue circle. Consult the radar reflectivity data in Fig. 8.4 and the storm total rainfall map you previously constructed.

2. Explain your answer to question 1.

3. Locate the signature of Hugo's eye and eyewall. Based on these and the distance reference provided in Fig. 8.12, calculate the propagation speed of Hugo for each time interval. What affect does landfall have on the speed?

4. How does Hugo's precipitation structure change after landfall?

5. Hurricane winds and storm surge are greatest in the eyewall on the right side of the storm when looking the direction of storm motion. Based on the radar intensity data where do you expect the greatest damage from these hazards occurred with Hugo? Mark the spot on Fig. 8.13 with a red "X".

Lab 44: Wind Generated Waves

Figure 8.14 Large breaker along Ka'iwi Coast on Oahu poses a tangible hazard to local fishermen and tourists.

INTRODUCTION

Hawai'i is famous for it's large surf. People travel to Hawai'i from around the world to experience the islands' reef and shore breaks. Yet many tourists and locals alike are not aware of the dangers high surf presents. High surf is the number one weather-related killer in Hawai'i. The hazards include dangerous shorebreak, strong rip tides, and coastal inundation. High surf also greatly affects marine transportation to and from the islands, as well as fishing, diving, sailing and other forms of recreation. A good understanding of the origin and evolution of waves and swell will help people living in Hawai'i safely enjoy our seas and shores.

The highest wind-produced wave ever recorded was 34 meters (112 feet) high. In 1933, the U.S. Navy tanker, *U.S.S. Ramapo*, was traveling from the Philippines to San Diego. During transit, the tanker encountered a storm system that produced strong winds (up to 70 mph) for more than a week across the Pacific Ocean. At the height of the storm, observers on the ship saw a mountainous wave that crested even with a platform on the crow's nest mast. From the height of the tanker's bridge, mast and stern, it was calculated that the wave was 34 meters high--as high as a nine-story building. Such immense waves are sometimes referred to as "rogue" waves and are implicated in the sinking of numerous ships, large and small.

WAVE DEVELOPMENT

When a winter storm develops over the ocean, all the ingredients needed to generate large waves are present; strong winds blowing over a long stretch of open deep water, and lasting for a long time. Waves begin as ripples and spread out in the direction of the wind. These ripples gain size as the winds continue to blow over the developing waves. Most of the waves that reach us here in Hawai'i in the middle of the Pacific Ocean originate from high winds associated with storms far from land. These storms provide the three main ingredients needed to make a wave. The first ingredient is **wind strength**. The wind must be strong enough to stir the ocean surface. The friction between the moving wind and the ocean's surface whisks up ripples, called capillary waves. These ripples increase in size as the wind continues to blow, creating wind waves and seas (Figure 8.15). The stronger the wind, the larger the resulting waves will be. **Wind duration**, or the length of time the wind is blowing, is the second ingredient. If the wind blows for a long time, a large amount of energy is transferred to the ocean, making waves bigger. Storms that move slower will have more time to transfer energy to the ocean, making larger waves. **Wind fetch** is the third necessary factor. The fetch is the distance over which the wind blows without changing direction. As you would expect, large storms cover a long distance, and produce large waves. Increasing the wind strength, wind duration, or wind fetch will result in larger waves.

While only these three factors are necessary to create a wave, two more contribute to wave height. **Air-sea temperature difference** is the first. If the ocean surface is warmer than the air above it, warm air near the surface will become unstable and tend to rise and will be replaced by cooler air and greater wind speeds from above the surface in the unstable region. Warm ocean waters contribute to convection and storm development. **Ocean depth** is the second contributing factor. As waves approach shore, eventually the wave bottom will encounter the sea floor.

259

This happens when the ocean depth equals half of the wavelength. The wave continues toward the shore as ocean depth decreases. The wave bottom drags on the seafloor and slows down. The top, however, continues on at a slightly higher speed. Water builds up behind, increasing the wave's height.

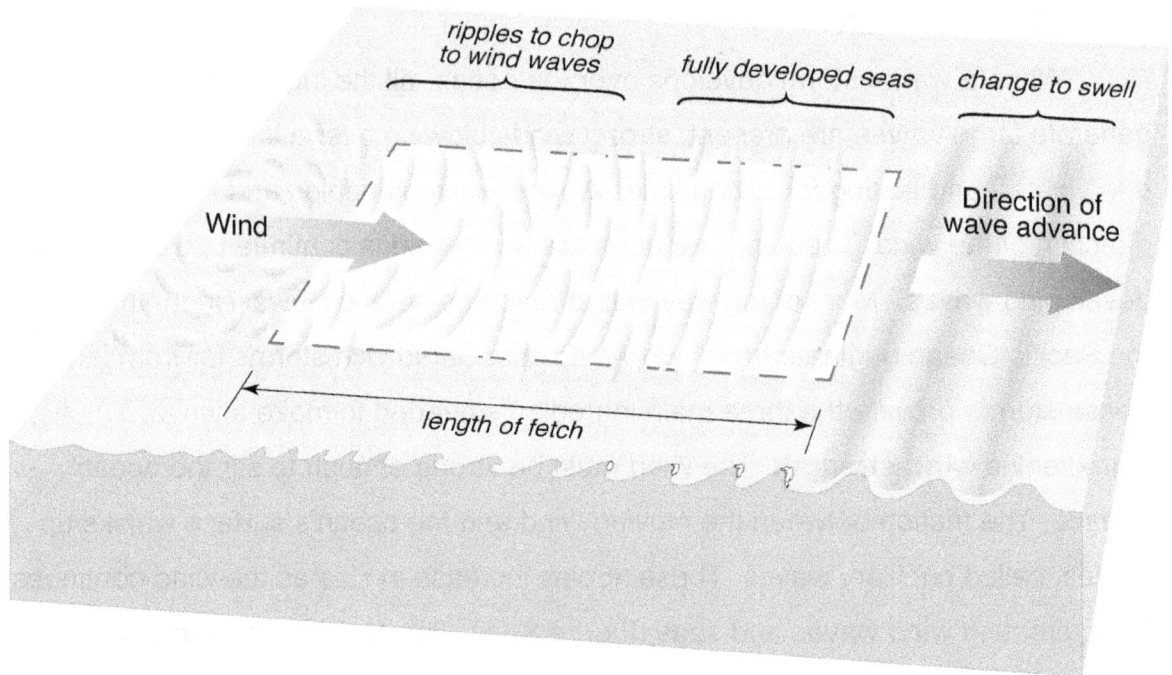

Figure 8.15 Schematic showing a wind fetch and the development of wind wave, seas, and swell.

WAVE LIFECYLCE & DECAY

When waves travel outside of the region of high winds they gradually transform from an unorganized jumble of steep waves into orderly swell. How does this happen? Waves with similar wavelengths travel at similar speeds, and therefore organize themselves into groups of waves called **swells**. Swells with greater wavelengths travel at higher speeds. This sorting of waves by wave length is called as **dispersion**. The speed of the swell is called the **group velocity**. The group velocity is exactly half of the individual wave speed. The reason for the slower group velocity is that the lead wave in the group must lift the ocean surface, expending energy in the process, causing it to decay. Meanwhile the trailing wave gains energy from the waves in front and grows in amplitude. Thus, the wave train looses energy at the front and gains energy at the back and the group travels as a slower speed then the individual waves that make up the group.

Some types of waves, like visible light waves, only spread over a small angle as they travel. Think of a flashlight. When we turn on the flashlight, we see a beam of light that more or less travels in the direction we point the flashlight. Ocean waves, however, spread out more as they travel. In particular, young steeper waves spread more than older less-steep waves. This spreading out of the wave energy decreases wave height across the swell. In open ocean, the largest waves propagate in the same direction the original wind blew, with decreasing swell height to either side, creating a fan-shaped swell. This effect is called **angular spreading**.

So far, we have discussed the propagation of waves from one storm travelling unaffected by other winds and swells. At sea, however, this rarely, if ever, occurs. The ocean is a complicated canvas, with countless swells, winds, and shores affecting wave propagation. If a swell encounters wind or another swell moving in the same direction, wave energy and height increase. Conversely, if a swell meets wind or swell moving in the opposite direction, wave energy and height decrease. If the opposing force is strong enough, waves can disappear entirely. The decrease in wave energy as is travels through the ocean is **dissipation**.

Once a wave nears shore, the wave bottom encounters the seafloor, creating friction that drains energy from the wave, or **shoaling**. This occurs when ocean depth is equal to one half of the wavelength. The trough is forced to slow down, while the crest pushes forward at higher speeds. Some parts of the swell slow down more than others, due to variations in ocean depth. This **refracts** the waves, or bends the waves towards the shore. The swell line begins to match the shoreline, so that waves reach points of land that stick out furthest first.

Eventually the difference in speeds of the trough and crest causes the crest to lurch forward. At this point the wave is known as a breaker, and is recognized by the crest plunging forward into a "lip", or curtain of water that heaves toward the shore. This creates a hollow region in the wave, known as the "barrel". The lip carries much of the wave's energy, and can have dangerous consequences for the shoreline and anyone on it.

RIP CURRENTS

Waves break on the shore continuously pushing water toward the shore. To avoid piling up, the water must find a route back to sea. This is acheived through narrow jets of strong current rushing away from the shore, called **rip currents**. Rip currents can also form as a result of strong onshore winds, especially along windward coasts. Rip currents tend to form in areas where there is a break in the reef caused, for example, by an ancient stream bed, where the water depth is greater. It can be difficult to spot rip currents from the shore, as their appearance varies depending on the shoreline. They pose a great danger to swimmers, divers, and surfers alike, as because of the speed of the offshore currents. If caught in a rip current, the best way to escape is to swim diagonally toward shore. Do not attempt to swim directly against the current.

Figure 8.16 Example of rip current along shore. The break in the white water indicates the jet of strong current flowing oceanward.

WAVE CLIMATOLOGY IN HAWAI'I

Hawai'i experiences swells from different directions depending on the season. The Pacific storm track exerts a strong influence on Hawai'i's swells. This is because strong storms are generated primarily in the winter hemisphere, where temperature differences from poles to tropics are much larger than in the summer hemisphere. The stronger temperature and pressure gradients create storms and strong winds over the open ocean, which are felt at the shore as heightened swells.

In the Northern Hemisphere's winter season, these swells are generated to the north and northwest of Hawai'i. Accordingly, from November through March Hawai'i experiences world-famous north and northwest swells. The pattern of winter storms changes from year to year, especially when a strong El Niño or La Niña occurs (see Figure 8.17). Note how the center of fetch activity shifts slightly eastward when comparing the fetch pattern for El Niño to that of La Niña winters. When a storm moves in the same direction as the swell are propagating, the storm generates a captured fetch and extra large waves result. The largest wave events seen in Fig. 8.18 are associated with large storms that move toward Hawaii, causing high winds to move with the large swell in a captured fetch. The data in Fig. 8.18 was collected by professional life guards in Hawaii, using the method described in the Lab Activity, Estimating Breaker Heights.

In the Southern Hemisphere's winter season (May through September), when strong swells are generated south and southeast of Hawaii, we experience relatively large swells on the south- and southeast-facing shores.

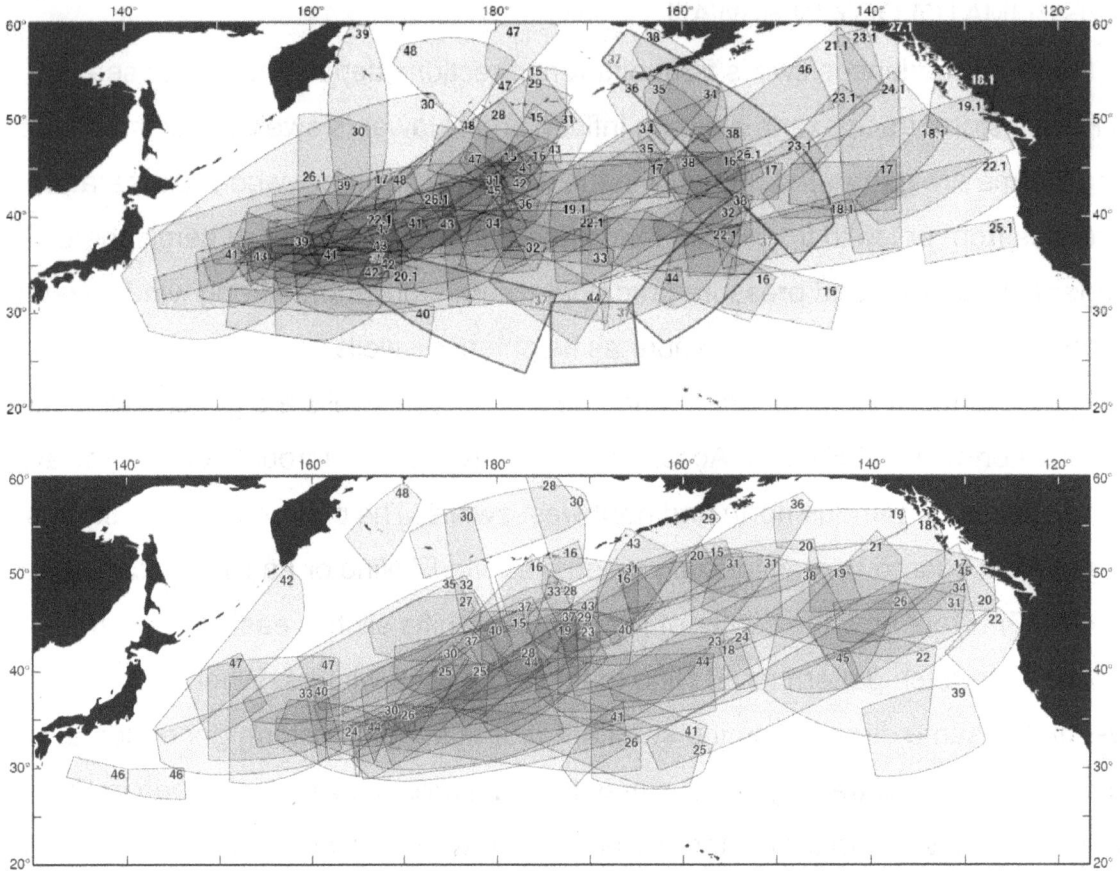

Figure 8.17 a) Fetches during the 2006-2007 El Niño event. The black outlined fetches are from a storm on 28-29 January 2007 that sent very large swells to both Hawaii and Alaska. b) Fetches during the 2007-2008 La Niña event.

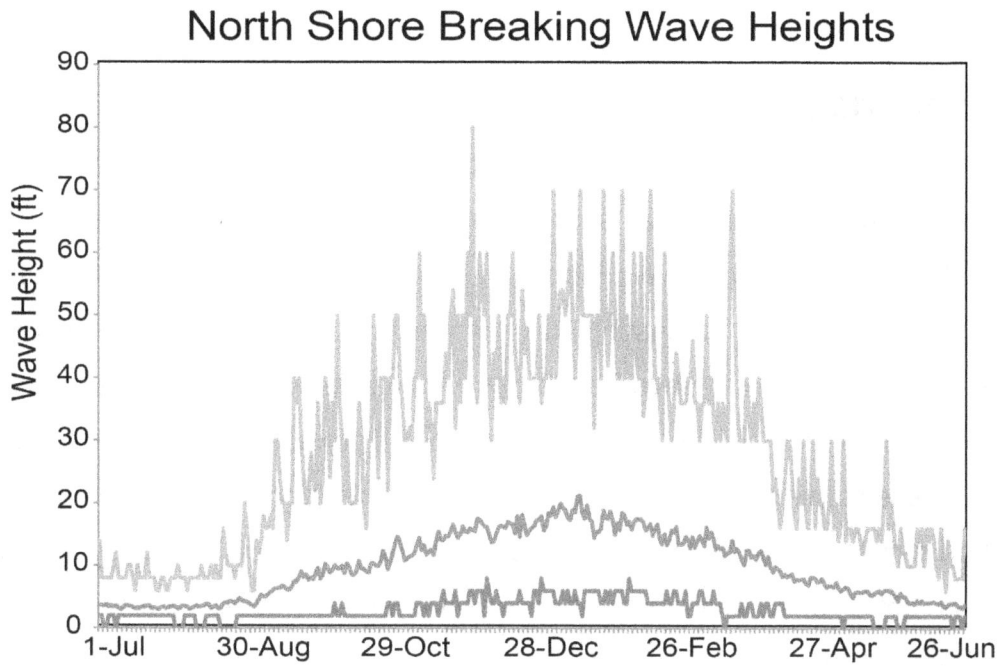

North Shore Breaking Wave Heights

Figure 8.18 Climatology of waves impacting Oahu's North Shore. Graph shows minimum, average, and maximum wave events as the function of time of year (graphic provided by Kelly Lance using data provided by Pat Caldwell).

Lab 44 Estimating Breaker Heights

INTRODUCTION

Measuring the height of waves breaking near the shore can be difficult, because waves are scale invariant, that is large waves and small waves look alike. To estimate the height of the wave face from shore, a surfer riding the wave is commonly used for scale. When forecasting surf, the National Weather Service (NWS) officially forecasts the height of a wave front or face from trough to peak as observed from shore just at the time of first breaking. In Hawaiʻi, an alternative approach to gaging wave size is commonly used by local surfers, some life guards, and beachgoers. This method estimates the height of the back of the wave, which is roughly one-half the wave height as seen from shore, therefore it underestimates the size of the waves by ~50%, causing some confusion.

During excessively large swells, the NWS will issue advisories and warnings to the public as necessary. Advisories are a way to let people know that waves are exceeding a safe height, and warnings are issued when there is an immediate threat to life and property. The height considered safe depends on which direction the shore faces. For Oʻahu's South and East facing shores, advisories are issued at 8 feet; for west facing shores, 12 feet; for north facing shores, 15 feet.

Figure 8.19 Large winter-storm generated wave crashing on Kaena Point, Oahu.

266

When water depth becomes less than half the wavelength, the waves "feel bottom." At this point the height increases and the wavelength and velocity decrease, but the period remains the same. The wave speed under these "shallow" water conditions is given by:

$$C = (gL)^{1/2}$$

where C is swell speed in shallow water, L is wavelength in meters, and g is gravity. Buoy data give the period (P) in seconds and the amplitude (A) in feet of open-ocean swell. Forecasters have a rule of thumb that states that the height (H) in feet of the waves observed at the beach can be estimated from the following formula:

$$H \sim .1226 \times P \times A$$

ACTIVITY

OBJECTIVE: The purpose of this lab is to learn to estimate the height of breaking waves when observed from the beach. So if you live near a surfing beach, go there and take your own photos and analyze them for this lab.

Assuming that a surfer is about 5 feet when slightly crouched, try to estimate the heights of the following five waves in Figs. 8.20 through 8.24. Keep in mind that the wave is measured as the peak wave face, just as it first begins to break.

Fig. 8.20 How large is this wave?

Fig. 8.21 How large is this wave?

Fig. 8.22 How large is this wave?

Fig. 8.23 How large is this wave?

Fig. 8.24 How large is this wave?

QUESTIONS:

1. What are your five wave-face estimates?

2. Describe any challenges you encountered in making your estimates.

3. From the sketch use the physical dimensions of the ship to calculate the height of the largest wave ever sighted, as the crew did. The *U.S.S. Ramapo* had a length of 164 meters. The crow's nest is located midship with a height of 17.5 meters. The wave had a period (P) of 14.8 seconds.

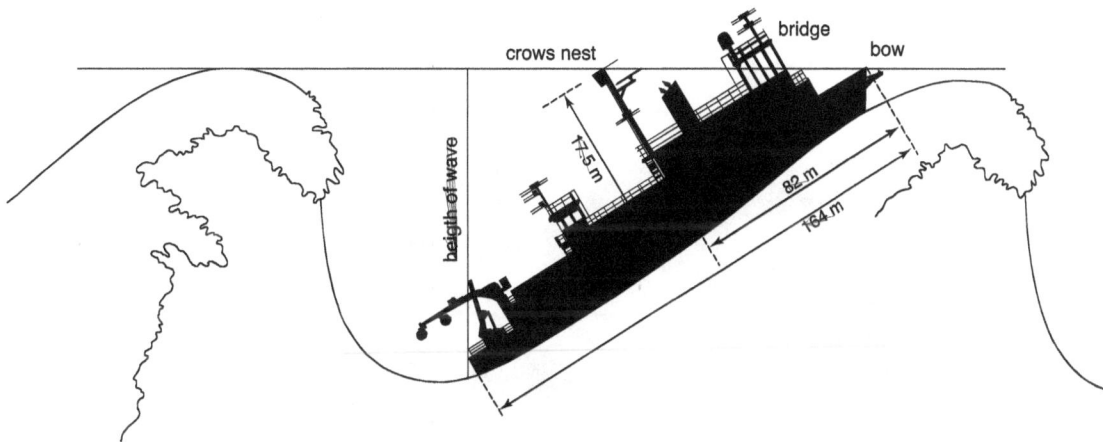

4. What is the length of the wave?
   [Use the formula: L (m) = 1.56 (m/s^2) x P^2(s^2).]

5. Using this value for the wavelength L and length of the ship, how can you determine the height of the wave? The observer is on the bridge.

Chapter 9 Atmospheric Pollution and Global Warming

'Man did not weave the web of life, he is merely a strand in it.
Whatever he does to the web, he does to himself.' Chief Seattle

Air pollution is composed of airborne substances (either solid, liquid, or gaseous) that threaten the health of the Earth's flora and fauna (including animals and people). The source of air pollution can be natural (forest fires, volcanic eruptions, wind born dust, etc.) or through human activities (burning of fossil fuels, release of refrigerants, etc.). By far the majority of air pollution is released into the atmosphere at the Earth's surface. Therefore, in the absence of winds and upward mixing, pollution can rapidly accumulate, especially in urban areas, leading to pollution episodes. The term "smog," a combination of the words smoke and fog, was coined to describe such events. Cold, stable air drains into basins and valleys from adjacent higher terrain, especially during the longer nights of winter, causing suppressed air circulation in these regions. In December 1952, a particularly bad smog episode in London claimed the lives of over 4000 people. Since that time legislation, such as the Clean Air Act in the U.S., has reduced air pollution levels in urban areas, decreasing the threat of such disasters in developed nations.

A major pollutant of city air is carbon monoxide, a colorless, odorless gas emitted by internal combustion engines (primarily cars and trucks). Carbon monoxide is actively taken up by our respiratory systems displacing oxygen intake and leading to suffocation when the concentrations are too high. There has been about a 40% decrease in carbon monoxide in the U.S. since 1970, due to stricter air quality standards and the application of emission-control equipment.

Rain and cloud water has become more acidic as a result of the burning of fossil fuels that contain sulfur dioxide (SO_2) and nitrous oxides (NO_X). These pollutants recombine with oxygen molecules in cloud droplets to form sulfuric and nitric acids. These newly formed acids are flushed from the atmosphere through precipitation in the form of rain, snow, sleet, or fog. The highest concentration of acid is found in cloud water that has not yet been diluted. In mountainous regions, such as the Appalachian Mountains, acid fogs have weakened many trees, making them susceptible to drought, disease, and insect infestation.

A primary constituent of contemporary urban smog episodes is ozone (O_3), a poisonous substance that irritates the eyes and respiratory system, and retards plant growth. In the lower atmosphere ozone forms in a chemical reaction that involves nitrogen dioxide (NO_2) and molecular oxygen (O_2) in the presence of ultraviolet light from the sun (thus the term photochemical smog). Ozone concentrations in the lower atmosphere tend to display a daily afternoon maximum and early morning minimum.

Ironically, ozone that occurs naturally in the stratosphere absorbs high-energy ultraviolet light in the upper atmosphere, thus blocking these harmful rays from reaching the Earth's surface. A decrease in the concentrations of stratospheric ozone globally (Fig. 9.1) due to a complex chemical interaction involving chlorofluorocarbons

(CFC) that depletes stratospheric ozone has raised concerns of increased incidence of skin cancer and crop damage .

Halley Bay Station
Antarctica

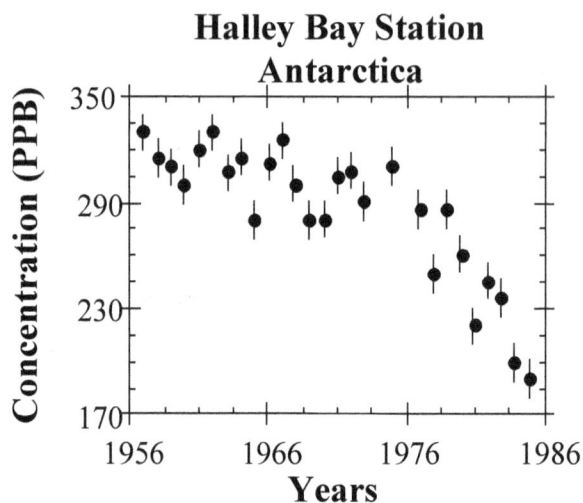

Figure 9.1 Ozone concentration in parts per billion (PPB) for Halley Bay, Antarctica.

The Clean Air Act and its amendments address mainly urban pollution and acid rain and include strategies for controlling high ozone levels in the lower atmosphere. It also offers protection of the stratospheric ozone layer by including provisions for recycling CFCs and calling for a ban on the production of CFCs.

Global Warming

The global climate of the Earth is largely determined by the balance between available (not counting that reflected back to space by clouds and the surface) incoming short wave (visible) solar radiation and outgoing long wave (infrared) radiation emitted to space by the Earth and atmosphere. Climate change depends upon changes that affect this radiation balance. Such changes can be separated into several categories:

1. Changes that affect the reflectivity of the Earth-atmosphere system for incoming short wave radiation:

i) changes in cloud cover (strong radiator in infrared and strong reflector in visible)

ii) changes in snow cover (strong radiator in infrared and strong reflector in visible)

iii) changes in the surface characteristics, e.g., deforestation, overgrazing, urbanization

iv) changes in aerosols, e.g., smoke, ash

2. Changes in the composition of the atmosphere affecting outgoing long wave radiation:

i) changes in radiatively active gases (e.g., water vapor, CO_2, methane, chlorofluorocarbons, etc.)

ii) changes in aerosols, e.g., smoke, ash

3. Astronomical changes affecting incoming solar radiation:

i) celestial mechanics of Earth's orbit

ii) solar aging

iii) Changes in the concentration of intergalactic dust

The complexity of the climate system is evident in the above outline and our knowledge of the complex interactions between the atmosphere, ocean, cryosphere (ice) and biosphere (plant/animal) systems is incomplete. Our ability to adequately observe the global climate is also insufficient. NASA's Mission to planet Earth and NOAA's Global Change Program include efforts to address these pressing issues.

Concern for global warming stems from the observed increase in gases such as carbon dioxide and methane, which inhibit outgoing long wave radiation. Records of temperature and carbon dioxide concentration derived from glacier ice cores show a strong relationship between these two. Under conditions of global warming, it is anticipated that the troposphere will warm, the stratosphere will cool, and ocean surface temperatures will rise, leading to more intense storm systems and wider swings in local climate, including enhanced droughts and flood events. Melting of ice sheets and warming of ocean temperatures produce sea level rise, with obvious implication for coastal areas. Observations, though incomplete, seem to support the concern.

Thinking Ahead

According to a NASA blue print, we can decrease CO_2 emissions over the next 30 years by 60% by doing the following five things:

1. Increase auto fuel efficiency by 40%
2. Use LED bulbs
3. Produce more efficient electric motors and better batteries
4. Apply tougher standards for refrigerators and dishwashers
5. Develop alternative renewable energy sources (e.g., wind, solar, wave, geothermal)

While human activity (burning of fossil fuels, agricultural activity, etc.) has increased the concentration of carbon dioxide and other radiatively active gases in the atmosphere since 1900, it has simultaneously increased the burden of aerosol, decreasing the amount of sunlight reaching the Earth's surface. Recent research has shown that areas where the aerosol input to the atmosphere is greatest have experienced cooler surface temperatures regionally. This effect may partially offset global warming and help explain why observed indications of global warming have been less than those forecast by global climate models.

Lab 45: Exhausting Problems

INTRODUCTION

For many high school students, one of the first signs of adulthood is being able to operate an automobile. But along with this wonderful freedom comes much responsibility. In your Driver's Education classes you learned a great deal about traffic safety, but what is the environmental impact of operating an automobile? How much does your automobile contribute to our air pollution problems?

In the introduction to this chapter a number of serious pollution problems directly related to the exhaust and operation of motor vehicles were mentioned. To summarize here, automobile exhaust contributes to global warming through the emission of carbon dioxide; to acid rain and smog through the emission of nitrous oxides, carbon monoxide and unburned hydrocarbons; and to stratospheric ozone depletion through leakage of chlorofluorocarbons from air conditioning units.

ACTIVITY

OBJECTIVE: In this activity, students will analyze the impact of their motor vehicle habits and the impact of pollution control devices.

PROCEDURE:

1. Fit a clean white sweat sock snugly over the cool tailpipe of your car, using a broad rubber band to keep it in place. Then turn on your ignition and let the car run for three minutes. Turn off the engine and carefully remove the sock. Inspect the sock for particulate matter.

2. Repeat the instructions in 1) with a second clean sock and a car that does not have emission control devices (pre-1975). If your car has no pollution control devices, choose a second car that does for this part.

3. Compute your annual fuel consumption by completing the following calculations.

Number of miles driven per week (average = 200)

a) _____

Estimated fuel mileage of your car (average = 20 mpg)

b) _____

Divide line a) by line b).

This is your weekly fuel consumption in gallons.

c) _____

Multiply line c) by 52. This is your annual fuel consumption in gallons.

d) _____

      We know that gasoline is actually a very complex mixture of chemicals that when burned releases another complex mixture of gases. In one gallon of gasoline, there is the following amount of air pollutants. For each pollutant in the data table below, use your annual fuel consumption figure (line d) to calculate your annual release of pollutants from your automobile.

Data Table

POLLUTANT	POUNDS PER GAL	X	ANNUAL GALLONS (line d)	=	MY ANNUAL RELEASE (in lbs)
CO_2	20.000	X		=	
NO_2	0.110	X		=	
CO	2.300	X		=	
Hydrocarbons	0.200	X		=	
Aldehydes	0.004	X		=	
Particulates	0.012	X		=	
Organic acids	0.004	X		=	
SO_2	0.009	X		=	
TOTAL	22.639	X	_____	=	_____

QUESTIONS:

1. Describe the difference between the appearances of the two socks? Can you estimate the percentage increase in the particulate pollution?

2. Compare the socks in your class. Can you make a statement about the appearance of the socks in the class on the basis of the age of the cars, their fuel efficiency, etc.?

3. Based on your calculations above, is your contribution to air pollution significant?

4. There are over 140 million automobiles in the United States. What would be the total amount of air pollution from automobiles if everyone drove the same as you?

5. Describe four ways in which you could reduce your automobile air pollution.

   a.

   b.

   c.

   d.

Lab 46: Understanding Acidity

INTRODUCTION

We have all heard the term "acid rain," but do we really understand what it is and how it affects the environment? In order to understand acid rain it is necessary to understand the pH scale. The pH scale was developed in 1909 by a Danish biochemist. The pH scale measures the number of hydrogen ions in an aqueous solution and determines the acidity or alkalinity of a solution. Values on this scale range from 0 to 14. A solution with a pH of 7 is neutral; acidic solutions have pH values below 7 and basic solutions have pH values above 7. The pH scale is logarithmic, which means that a change in pH by one number actually represents a change by a tenfold. For example, a solution with a pH of 5 is ten times more acidic than a solution with a pH of 6 and one hundred times more acidic than a solution with a pH of 7.

Very few substances are neutral (have a pH of 7). Distilled water is considered neutral. Background rainwater falling from clean air is not neutral; it has a pH value of 5.6. This is a result of the presence of carbon dioxide in the atmosphere, and the formation of carbonic acid in cloud water. Acid rain is defined as a solution with a pH less than that of background rainwater. Typically, pH values for acid rain range from 4.0 to 4.6. However cloud water with a pH as low as 2.3 has been observed in fogs in the Appalachian Mountains.

ACTIVITY

OBJECTIVE: The purpose of this activity is to investigate the pH of various acids and bases.

MATERIALS:

§ aspirin   § milk- fresh

§ sugar   § milk- sour

§ cocoa   § vinegar

§ indigestion tablets   § rainwater

§ salt   § orange/ Lemon/ Apple juice

§ soap   § soda

§ tea/ coffee   § distilled water

§ pH paper   § tweezers

§ small cups   § stirring rod

PROCEDURE:

1. Add one teaspoon of each solid to one-quarter cup of distilled water and stir until dissolved. For any of the given tablets, add one tablet to one-quarter cup of water.

2. Use tweezers to dip the pH paper into each solution. Do not re-use the pH paper.

3. Determine the pH of each solution based upon the color scale. Record your observations and results in the table provided. Be sure to label each substance as an acid, a base or neutral.

QUESTIONS:

1. Which solution was the most acidic? Basic?

2. What would happen to the pH value if you dilute the liquid solutions with distilled water?

3. What would have happened to the acidity of the solution if the solids had not been dissolved in water?

4. What would happen to the pH of a solution if it were left uncovered for a weekend?

5. Why are the lowest values of pH observed in cloud water rather than in rainwater? What are the consequences of that observation to alpine forests that exist at an altitude above cloud base?

Data Table

Substance	pH Value	Acid or Base	Observations
Aspirin			
Sugar			
Cocoa			
Indigestion Tablets			
Salt			
Soap			
Tea/ Coffee			
Milk- fresh			
Milk-Sour			
Vinegar			
Rain water			
Juice			
Soda			

Lab 47: Monitoring Acid Rain

INTRODUCTION

The acidity of rain water depends upon a combination of naturally occurring acids in the atmosphere and acids that are the result of man made pollutants introduced into the atmosphere from a variety of sources. Three types of acid commonly found in rainwater include carbonic acid, nitric acid, and sulfuric acid. The concentration of carbonic acid depends upon the presence of carbon dioxide; and gives rainwater its natural background acidity (pH = 5.6). Sulfuric and nitric acids are the product of sulfur dioxides and nitric oxides introduced into the atmosphere by industrial power plants and motor vehicles, respectively. *Check the local pH index, if one is available for the area.*

ACTIVITY

OBJECTIVES: The goals of this activity are to learn how to take pH measurements of liquid precipitation under field conditions, and to gain knowledge of the variation in acid rain episodes, analyze the human impacts on the environment, and become aware that the degree of accuracy of measurements has to be taken into account when results are interpreted.

MATERIALS:

§ pencil, greenhouse type

§ 2 pocket type, student thermometers, single or double scale

§ heavy duty, 1 m2 plastic tarp

§ 500 mL, plastic graduated cylinder

§ plastic rain gauge to measure up to 15 cm. of rain

§ 10+ pH sticks, wide range (permanent type, ColorpHast)

§ 18+pH sticks, narrow range (permanent type, ColorpHast)

§ storage bag/vial to store used, dry pH papers, (color is permanent)

§ waterproof plastic case

§ plastic laminated pH Colorchart, wide range

§ plastic laminated pH Colorchart, narrow range

§ map of expedition area

PROCEDURE:

1. Before or at the first hint of rain, spread the 1 m² plastic tarp on the ground, in an open area, away from trees, and anchor it.

2. Position the rain collecting graduated cylinder so that it catches rainwater run-off from the tarp. This will require some ingenuity and you may have to rearrange your anchors. Keep tarp clean.

3. Place the rain gauge in the same general area as the tarp.

4. At intervals throughout the storm, collect water samples and record the sample data.

5. Record the following information on a copy of the General Data Sheet that follows:

   A. Record name or names or members in observation group

   B. Date sample taken.

   C. Describe the rain sample site location and vegetation. Provide as much detail as you can.

   D. Record actual location with map coordinates or attach a map to the Data Sheet with the sample site location clearly marked.

   E. Note the duration of rainfall. Time Began, Time Stopped, and Total Time of Precipitation.

   F. Describe the cloud type if you can or describe sky conditions: completely overcast, thunderstorms, passing showers, etc., or use a word description of your own. You can check it against a weather guide when you return home.

6. If it rains long enough, attempt to collect at least three samples during the rainstorm's duration. Record the following data for each sample in the Data Table:

G. Time sample taken.

H. Rain intensity - i.e., downpour, light, medium, heavy, mist, etc.

I. Wind direction.

J. Wind speed (best estimate). (Beaufort scale)

K. Air temperature.

L. Amount of rainfall accumulated per sample (Rain Gauge).

M. Determine and record the pH. Use the wide range pH paper, which gives the pH to within one pH unit (Record). Obtain a more accurate pH reading by using the appropriate narrow range pH paper. Take three readings and record them on the Data Sheet.

N. Determine and record the overall pH of the entire rainfall by testing the rainwater collected in the rain gauge during the rain. Record.

O. Record total accumulation (Rain Gauge).

P. Dry the used pH papers, label them with the sample number, and place them in the used pH storage vial. Do not discard used pH papers, pack them out.

General Data   NAME_____

1.   Date Sample Taken _____

2.   Site Description

3.   Map Location:

     Coordinates _____
          (Or attach map with location marked)

4.   Rainfall Duration:

          Time Rain Began _____

          Time Rain Stopped _____

          Total Time of Precipitation _____

5.   Cloud Description/Sky Conditions
     _____

6.   pH of Total Accumulation _____

7.   Total Rain Accumulation (run-off) _____ cm.

8.   Total Rainfall (measured in rain gauge) _____ cm.

Data Table

Sample #	1	2	3	4	5
1. Time Sample Taken					
2. Rain Intensity					
3. Wind Direction					
4. Wind Speed					
5. Air Temperature					
6. Sample Accumulation					
7. pH of Sample					
a. Wide Range					
b. Narrow Range					
Trial 1					
Trial 2					
Trial 3					
8. Local pH Index (if available)					

QUESTIONS:

1. How did the acidity of the samples vary with location and wind direction. Can the observations be explained in the context of the location of pollution sources in the area?

2. How did the acidity of the samples vary with the time the sample was taken; early or late in the day, early or late in the rain event. Can these differences be explained in the context of diurnal variations in pollution or the impact of precipitation on the quality of the air?

3. Investigate the effects of soot/particulates from a gas engine's exhaust, placed on fabrics and moistened with distilled water.

5. How do the leaves of a plant react to the application acids of varying pH? Examine the tensile properties of yarn exposed to acid solutions of varying pH.

Lab 48: The Atmosphere Effect and Global Warming

INTRODUCTION

When visible light passes through large windows of a greenhouse, it is largely absorbed after striking objects within (and in the case of plants results in photosynthesis). These objects heat up and in turn conduct heat into the air of the greenhouse as *sensible heat* (heat you can *sense* with a thermometer). The glass or Plexiglas windows prevent the mounting sensible heat inside the greenhouse from escaping and mixing with cooler outside air and the building stays warm even on cold sunny days. Unlike a greenhouse, the atmosphere does not have a barrier that limits mixing of the air, such as the glass in a greenhouse, thus the term "greenhouse warming" when applied to the atmosphere is actually a misnomer.

How does the atmosphere work? Recall that light is emitted by all objects in wavelengths and energies that depend on the temperature of the object. As the Earth's surface is warmed by sunlight, it emits increasing amounts of infrared radiation, radiation with a longer wavelength than sunlight. Several gases in the atmosphere absorb and reradiate this energy back to the surface and are mislabeled "greenhouse gases." Examples include water vapor (H_2O), carbon dioxide (CO_2), and methane (CH_2).   Water vapor is the most important greenhouse gas in the Earth's atmosphere. The latter two gases are now increasing daily because of exhaust from cars and burning coal. At present, the atmosphere traps only a small part of the infrared radiation that is emitted into the air at the surface of Earth. As the amounts of carbon dioxide and methane increase, the Earth will become gradually warmer as the atmosphere absorbs and reradiates more and more of the infrared radiation that would otherwise have passed directly out to space.

The temperature here tonight on Earth depends upon several factors -- seasons, latitude, altitude, proximity to oceans, and prevailing weather patterns. But the temperature on some of our closest planet "neighbors" is thought by many to be simply a result of that planet's distance from the sun. As we hope to demonstrate in part two of this exercise, there is another important factor.

ACTIVITY

OBJECTIVE: The objective of this activity is to observe and to investigate the processes that occur in a greenhouse and compare them to those that take place in the Earth's atmosphere. Part two of this activity will compare and contrast the environments of the other terrestrial (Earth-like) planets in our solar system with each other and Earth.

MATERIALS:

For the class:

§ one large bag of potting soil

§ one box of plastic wrap

For each student or group of students:

§ one rubber band

§ two thermometers

§ two large disposable plastic cups

§ hole punch

PROCEDURE:

1. Use the hole punch to make a hole big enough for a thermometer to be inserted about an inch from the top of each plastic cup .

2. Fill each cup with dirt until the soil is about one inch below the hole just made.

3. Insert a thermometer through each hole so that the bulb is about one inch above the dirt and centered in the middle of the cup. **Caution: Do not force the thermometer through the hole. If it will not go, punch a bigger hole.**

4. Turn the thermometer so that it can be read.

5. Cover one cup with plastic wrap, and leave the other cup uncovered. Secure the plastic wrap on the cup with a rubber band as shown below in Fig. 9.2.

6. On a sunny day, take the two cups outside at the beginning of the class period, and place them where they will not be disturbed. Stabilize the thermometers so they will not move.

7. Record the initial temperature on each thermometer in the Data Table below.

8. Record temperatures every 5 minutes for 30 minutes.

9. Make a graph for the temperatures in each cup on the same sheet of graph paper. Graph the temperatures on the vertical scale and the time on the horizontal scale. Designate each line as being for the covered cup or the uncovered cup.

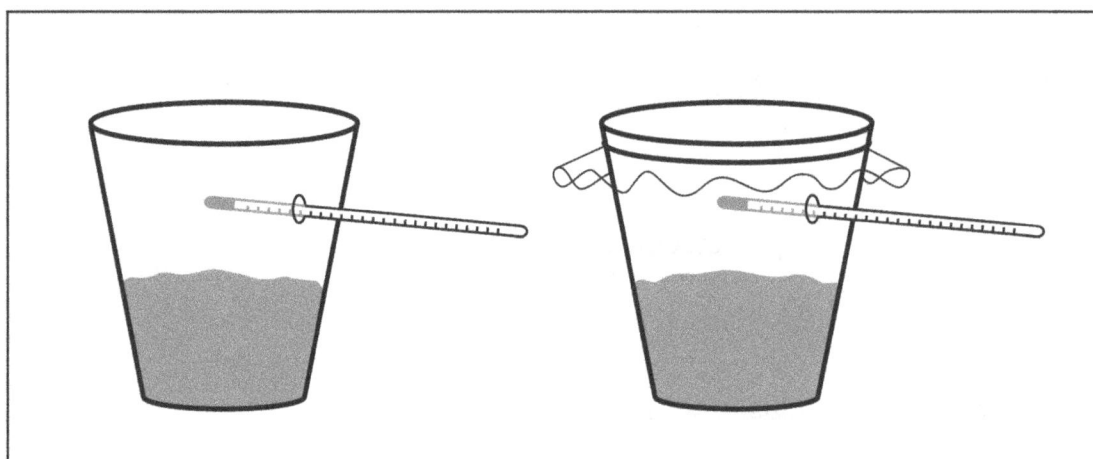

Figure 9.2 Schematic diagram

QUESTIONS:

1. Describe what happened to the temperature in each cup.

2. Given that infrared radiation will pass freely through the plastic wrap over the top of the cup, what is making the temperature rise more rapidly in the covered cup?

3. In a glass greenhouse the infrared rays emitted within are trapped to some extent by the glass, not so in a Plexiglas greenhouse, yet the latter still stays warm. What does this tell you about the importance of the infrared radiation in keeping a greenhouse warm?

4. In the absence of other factors, what will be the result for global surface air temperatures if greenhouse gases continue to increase?

5. Extra credit: What factors may help offset global warming?

6. Extra credit: Potting soil tends to be moist. How might the results of this experiment been different if dry soil or sand had been used instead? (See the introduction to Chapter 4 for a hint.)

Data Table

	COVERED CUP	UNCOVERED CUP
0 min.	°	°
5 min.	°	°
10 min.	°	°
15 min.	°	°
20 min.	°	°
25 min.	°	°
30 min.	°	°

PART II: In part two of this activity, we will examine the four terrestrial (Earth-like) planets -- Mercury, Venus, Mars and Earth. Table 9.1 is a chart of data concerning these planets. Please take a moment to review these data then answer the questions that follow.

Table 9.1 Temperature range and atmospheric composition for select planets in our solar system.

PLANET	DISTANCE FROM SUN (x1,000,000 KM)	LOWEST TEMP (°C)	HIGHEST TEMP (°C)	TEMP RANGE (°C)	ATMO-SPHERE
Mercury	58	-173	427	600	None
Venus	107	430	480	50	CO_2
Earth	149	-73	50	123	N_2, O_2
Mars	223	-125	-60	65	CO_2

QUESTIONS:

1.   Which planet has the coldest nighttime temperatures? Explain?

2.   Which planet has the highest daytime temperature? Explain?

3.   Which planet has the largest temperature range? Explain?

4.   Which planet has the smallest temperature range? Explain?

5.   What two factors do Venus and Mars have in common?
     a.

     b.

6. From the data above, what do you think the effect of a high CO_2 atmosphere is on temperatures observed?

8. What do you think would happen to the Earth's daily temperature cycle if we suddenly lost our atmosphere?

9. What do you think would happen to the Earth's global climate if the percentage of CO_2 in the Earth's atmosphere were to double over the next 30 years?

Lab 49: Volcanic Eruption

INTRODUCTION

Volcanoes are a large source of atmospheric pollution. Mt. Kilauea on the Island of Hawaii expels as much pollution in the form of sulfur gases into the atmosphere annually as the combined output of all coal-burning electric plants in the United States. Scientists are beginning to monitor this pollution source globally using satellite data. Catastrophic eruptions such as Mt. St. Helens in Washington state in 1980, and Mt. Pinatubo in the Philippines in 1991 can inject significant ash and sulfur gases into the stratosphere, where resulting fine particles can remain suspended for periods up to several years. These ash clouds block a portion of the incoming solar radiation from reaching the surface, resulting in cooling of the lower atmosphere. In the immediate proximity of a volcanic eruption the ash plume presents a distinct hazard to aircraft by causing jet engines to seize up after they ingest ash. The ash accumulates in a smooth blanket several inches deep downwind of the mountain (Fig. 9.3).

ACTIVITY

OBJECTIVE: The objective of this experiment is to observe and analyze the dispersion of ash from a model volcano.

MATERIALS:
§ ammonium dichromate; an orange-yellow crystalline compound
§ metal cookie tray or plate
§ match/lighter

PROCEDURE:

1. In a well-ventilated area outdoors, make a small pile of ammonium dichromate about 5-10 cm across on the metal cookie tray or plate. If there is no ambient wind consider employing a portable fan for this purpose.

2. Light top of the pile with a match or lighter and watch the ammonium dichromate gradually spark and expand. Although the burn temperature is relatively low, avoid standing too close to the demonstration to prevent inhalation of the ash. The demonstration lasts about 1 minute and provides a nice demonstration of a "strombolian" type volcanic eruption forming a "scoria" (cinder) cone. **Make careful observations of this process.**

3. Using a pencil or pointed stick make three lines or contours of approximately equal depth in the ash around the volcano. Choose your contour depths at equal increments (say 1 cm, 2 cm, 3 cm) to give a reasonable portrayal of the ash distribution.

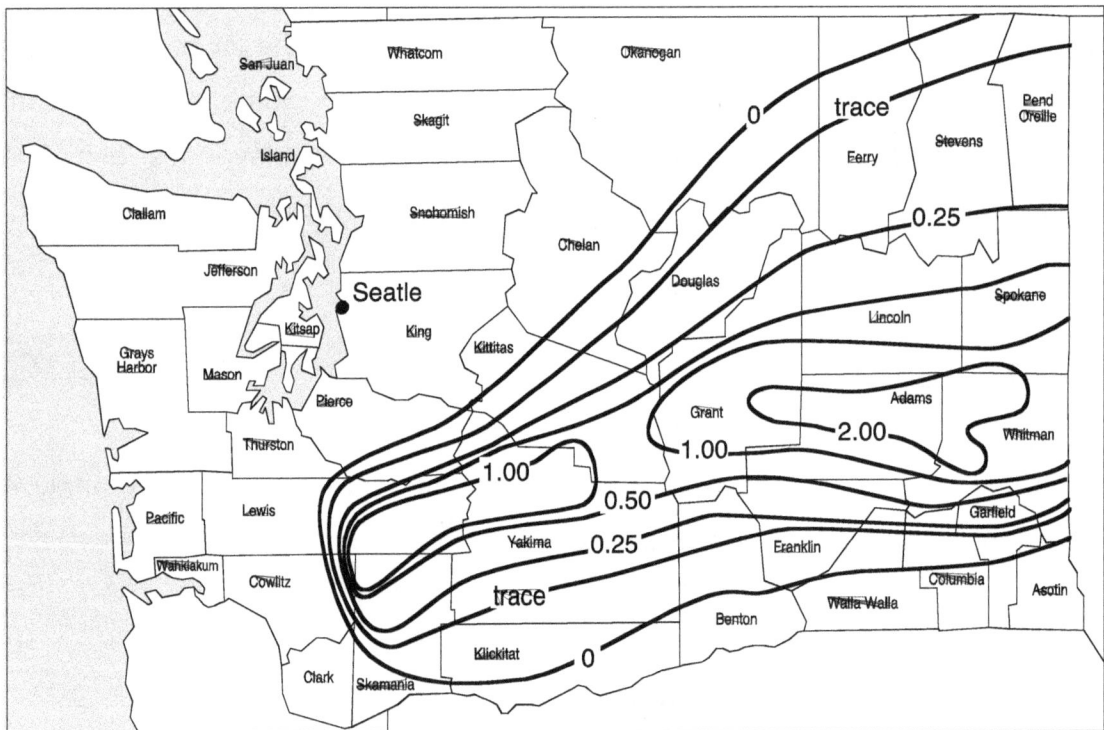

Figure 9.3 Contours of ash depth following the eruption of Mt. St. Helens on 18 May 1980. Up to three inches of ash covered large areas of eastern Washington.

QUESTIONS:

1. Carefully describe your observations during the demonstration. What was the ambient wind direction and speed?

2. How was the shape of the contours affected by the ambient wind? How do your contours differ from those in Fig. 9.3? Can you explain the differences?

3. Southerly winds are common over Washington state in the spring. While ash provides good fertilizer for fruit trees in eastern Washington, describe some of the hazards/problems that the ash cloud from St. Helens might have produced under southerly flow?

4. Make a sketch of your final volcano and the distribution of the resulting ash as found from the contours.

Chapter 10 Sky Lights

'There are only two ways to live your life. One is as though nothing is a miracle. The other is as though everything is a miracle.' Albert Einstein

Rainbows, halos, glories, and similar optical phenomena observed in the atmosphere provide for some of the most breathtaking spectacles in nature. It is not surprising that religious writings have interpreted the sighting of such phenomena as omens variously portending prosperity, death, or war. In no way diminishing their splendor, atmospheric optics are generally the result of a combination of one or more of the following four physical mechanisms: scattering, refraction, reflection, and diffraction.

Scattering is the process by which small particles diffuse a portion of the incident radiation in all directions. The amount of light that is scattered depends upon the size of the particle. Scattering of sunlight by air molecules is very sensitive to the wavelength of the light with the shorter blue wavelengths scattered almost four times more efficiently than the longer red wavelengths. This accounts for the blue sky we see. Scattering by air molecules is referred to as *Rayleigh scattering*, after Lord Rayleigh who first described it. In the absence of Rayleigh scattering the sky would appear black, as it does on the moon. As the sun approaches the horizon, the path length of the sunlight through the air increases and more and more of the blue light has been scattered away by the intervening air, thus the sun appears increasingly red.

For larger particles such as cloud droplets, haze, and smoke, the scattering falls into what is called the *Mie scattering* regime (named after Gustav Mie who developed the more general theory of light scattering) in which there is less dependence on wavelength, rendering the scattered light neutral or whitish in color. This regime accounts for the white appearance of clouds when the sun is high.

Another physical mechanism that results in atmospheric optical phenomena is *refraction.* Refraction is the bending of light as it passes from one medium to another (Figure 10.1). Refraction can also occur within a given medium when density differences cause the light to bend gradually. Refraction results from the fact that the speed of light varies according to the property of the medium through which it passes.

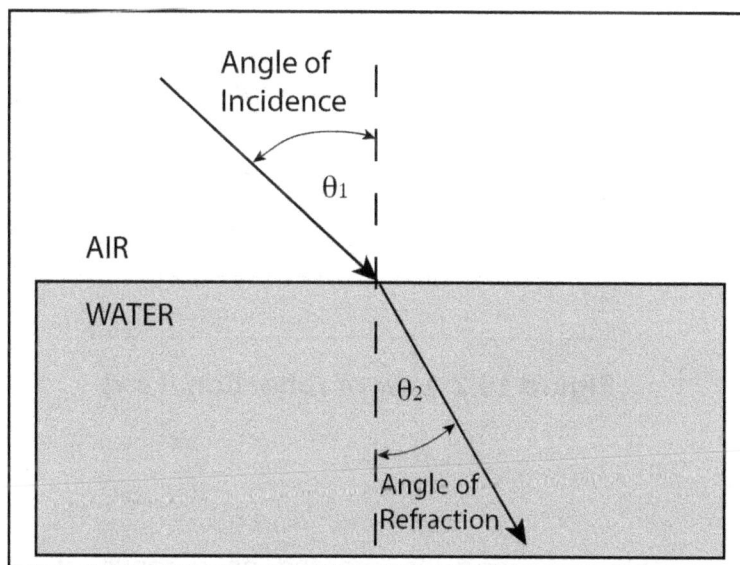

Figure 10.1 Refraction of sunlight by water

The amount of refraction depends upon the density of the material, the angle at which the light enters the material, and the wavelength of the light. As a result of the wavelength dependence, white light is separated into a spectrum of colors after it passes through a glass prism. Red light is refracted least due to its large wavelength, and blue light is refracted most due to its short wavelength.

A common physical mechanism that produces interesting optical effects is *reflection*. When reflection occurs an object or ray of light bounces off a surface at the same angle at which it strikes the surface. This is best stated in the *Law of Reflection*: the angle of incidence is equal to the angle of reflection (Figure 10.2). For instance, when light strikes a mirror, it is reflected in the same way a ball is reflected after striking a wall. Surfaces of discontinuity, such as air- water interfaces, result in a partial reflection of the incoming radiation.

The fourth physical mechanism mentioned is *diffraction*. Diffraction occurs when light bends along the boundary of an object. Diffracted light from many uniformly sized cloud droplets can interfere in such a way as to produce a pattern of colored light. A region of bright color is then the place where light waves of a certain wavelength have crests that arrive simultaneously. The rainbow colored pattern seen in modern compact disks is an example of diffraction.

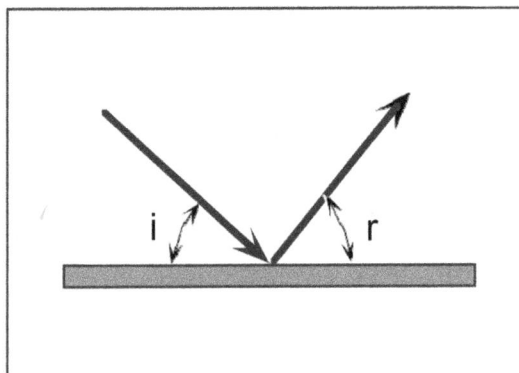

Figure 10.2 Law of reflection (i = r)

Rainbows

Rainbows are optical phenomena created as a result of both refraction and reflection of sunlight by raindrops. As sunlight enters a raindrop, it is refracted once then reflected off the backside of the droplet and refracted again before the light leaves the drop (Fig. 10.3) The result is known as a primary rainbow because light is reflected only once. Rainbows can be seen when an observer has his back to the sun and is facing illuminated rain.

As explained earlier, refraction separates light according to the wavelength of the light. When a rainbow is formed, the light with a longer wavelength (red) appears along the outer edge of the rainbow and the light with a shorter wavelength (blue) is along the inside. An acronym that is commonly used to remember the color order of a rainbow is ROY G. BIV (red, orange, yellow, green, blue, indigo and violet).

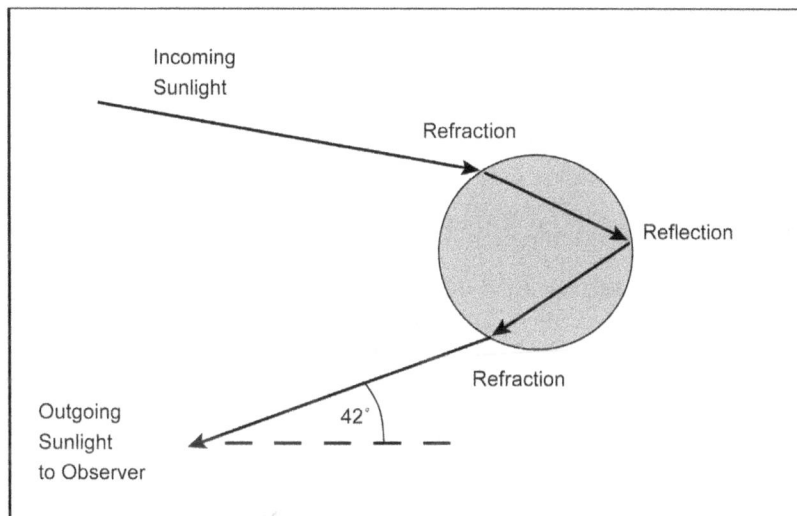

Figure 10.3 Primary rainbow

Often times a second, dimmer rainbow is observed next to the primary rainbow. This is known as a secondary rainbow because the light in the raindrops is reflected two times, instead of only once (Figure 10.4). The order of the colors of the secondary rainbow is reversed, with the blue on the outside and the red on the inside.

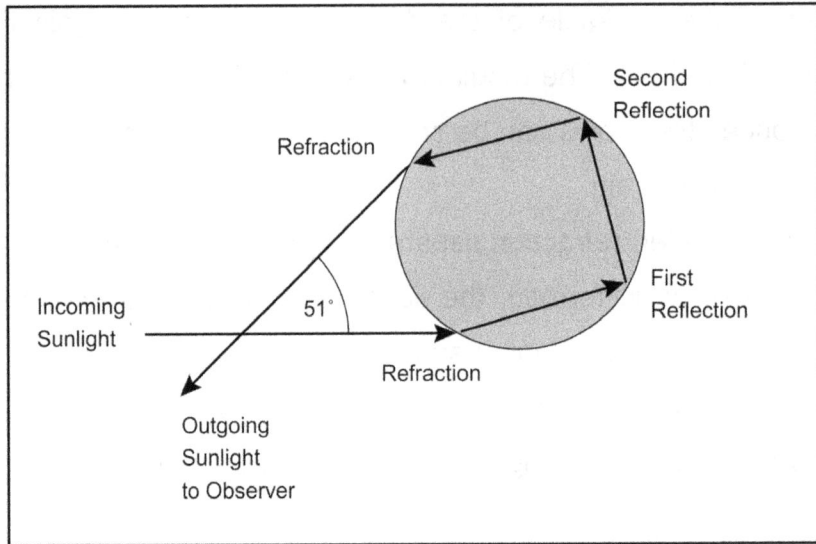

Figure 10.4 Secondary rainbow

An optical phenomenon that occurs frequently but is often overlooked is the *halo.* The word halo is a general term referring to any ring or arc that is created as a result of reflection and/or refraction of light by ice crystals. Most commonly, a halo forms around the sun or moon when light rays are refracted by ice crystals in cirrostratus clouds.

Halos are often observed as whitish rings around the sun or moon. A distinct rainbow spectrum is sometimes be observed around the sun. When color is present in a halo, red is closest to the sun, followed by yellow, green and blue on the outside. The colors of a halo are not usually as vivid as they are in a rainbow, and thus are often overlooked. The most common halo occurs at a radial distance of 22.5° from the sun, and is produced as a result of refraction of sunlight through the sides of hexagonal, pencil-shaped ice crystals that are oriented randomly.

The presence of altocumulus and cirrocumulus clouds composed of uniform sized cloud droplets allow light waves from the sun or moon to bend around them and thus create an interference pattern we call a *corona.* Corona are very common around the moon and are often confused with halos, but are actually the result of a very different physical mechanism, diffraction. The radius of a corona, typically only a few degrees, is much smaller than common halos. Colorful coronas around the sun are sometimes called Bishop's rings and differ from halos in their color order. Red is present on the outer edge, then fades to a blue or almost white along the inside. In good examples of Bishop's rings several concentric rainbow-colored rings can appear. The presence of color in a corona is the result of interference of light waves.

Glories are another type of atmospheric optical phenomenon that many people overlook. In order to view a glory, the observer must be in bright sunshine and above clouds or fog. Therefore, a common place for viewing a glory is from an airplane or mountain ridge. A glory appears as a ring of color located around the shadow of the observer and has the reverse color order of a corona. The glory forms through a complex combination of diffraction and reflection of light by the cloud by droplet. Cloud droplets that produce glories are of uniform size, as is the case for coronas.

Lab 50: Bending Light

INTRODUCTION

In Fig. 10.1, a beam of light strikes a water surface at an angle. At the interface between the water and the air, part of the beam is reflected and part enters the glass. Notice that the light entering the glass bends at an angle. This bending is called refraction, and it occurs whenever light crosses, at an angle, the boundary between two media of different densities.

In Fig. 10.1, light is traveling from a medium of low density to one of high density (air to glass). As the light enters the glass, it slows down and refracts toward normal. As n_2 (index of refraction) in the second medium increases, the angle θ_2 (as measured from the normal) decreases.

Willebrord Snell (1591-1626) is credited with the discovery of the relationships involved in refraction. He found that the change in velocity is analytically related to the angles formed at the boundary and the properties of the media through which the light is traveling. That is,

$$V_2/V_1 = sin\theta_2/sin\theta_1 = constant$$

V_1 = velocity of light in first medium
V_2 = velocity of light in second medium

The constant is a dimensionless number called the index of refraction and is represented by the letter "n." From the above equation, we get the widely used form of Snell's Law:

$$n_1 sin\theta_1 = n_2 sin\theta_2.$$

n_1 = index of refraction in first medium
n_2 = index of refraction in second medium

306

θ_1 = angle of incidence

θ_2 = angle of refraction

Snell's law provides the equation you will use to solve the problems in the following activity.

ACTIVITY

OBJECTIVE: The purpose of this activity is to draw the path a light ray follows as it travels through air, fresh water and salt water. You will then measure the angles formed and determine the index of refraction for fresh water and salt water.

MATERIALS:

§ glass or Plexiglas rectangular box

§ straight pins

§ cardboard

§ white paper

§ protractor with straight edge

§ salt water (50 grams of table salt to 1 liter of water)

PROCEDURE:

1. Secure a piece of paper to the cardboard and set it down on a table. Place the glass box on the center of the paper. Trace the outline of the box. Remove the box from the paper (see Fig. 10.5).

2. Find the midpoint on the back of the outline and mark its position (next to the line, not on it). Label this point B.

3. Draw a dotted line through Point B that is perpendicular to the outline of the box. This represents the normal to the back face of the box.

4. Draw another line, two centimeters in length that extends out from Point B at a 45-degree angle. Label the point at the end of this line "A" and the angle "θ_1".

5. Stick a straight pin into the paper at Point A and another one at Point B. Fill the glass box with fresh water and place it back onto the paper inside the outline.

6. From the front of the glass box, position yourself so that your line of sight is just above the table top. You should have a clear view of the pins on the other side. Move left or right until the pins at Points A and B appear evenly aligned. (You may close one eye, but do not tilt your head.)

7. Once aligned, mark a third point, "C," on the front side as close to the glass box as possible. Stick a straight pin into the paper at Point C. All three pins should appear to be aligned.

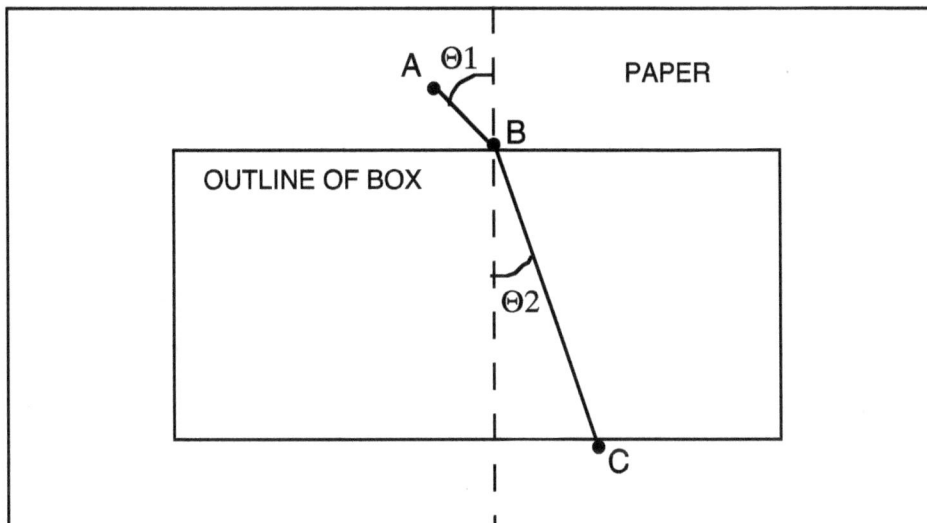

Figure 10.5 Schematic diagram

8. Carefully remove the glass box from the paper. Draw a line from Point B to Point C. Label the angle this line makes with the normal, "θ_2".

9. Measure the angle with a protractor.

10. Using Snell's Law, calculate the index of refraction for fresh water.

11. Fill the glass box with salt water and place it back over the outline. Repeat the steps above. Label your new point of alignment, "D," and the new angle, "θ_2."

12. Calculate the index of refraction for salt water.

QUESTIONS:

1. The accepted value for the index of refraction in fresh water is 1.33. How does your measured value compare with the accepted value? Express your answer in terms of percent error.

% Error = <u>(Difference between measured & accepted value) x 100</u>

accepted value

% Error = _____ %

2. Why is there no single accepted value for the index of refraction of seawater?

3. In the experiment, we did not take into account what happens to the light as it passes through the glass. In terms of speed and direction, how does the light behave as it travels from the air to the glass? From the glass to the water? From the water to the glass? And, finally, from the glass to the air again? Fill in the data table below.

Data Table

	SPEED	DIRECTION
air to glass:		
glass to water:		
water to glass:		
glass to air:		

Lab 51: Creating Rainbows

INTRODUCTION

It's always inspiring to see a rainbow after a rain shower. By now, most people realize that finding a pot of gold at the end of a rainbow is nothing more than a myth. Rainbows are the result of sunlight being refracted and reflected through water droplets. This activity will let you explore the formation of a rainbow and what helps or hinders the visibility of rainbows.

ACTIVITY

OBJECTIVE: The objective of this activity is to create and investigate a rainbow.

MATERIALS:

§ large glass beaker

§ water

§ cardboard

§ scissors

§ flashlight

§ milk

§ stirring rod

PROCEDURE:

1. Fill the glass beaker with water.
2. Cut a circle in the cardboard that is approximately the size of the beaker.
3. Take the beaker and cardboard into a dark room. Shine the flashlight through the hole in the cardboard into the beaker (See Fig. 10.6). Record your observations.
4. Vary the distance between the cardboard and the beaker. Record your observations.

5. Add a few drops of milk (can use non-fat dry milk or coffee creamer for this purpose) to the beaker and stir well. Repeat the above procedure. Record any changes in your observations.

6. With the milk added to the water, view the light through the beaker by looking opposite the light source. Record your observations.

QUESTIONS:

1. Why did a rainbow appear on the cardboard?

2. How did the distance between the beaker and the cardboard effect the image? Why is this so?

3. After the milk was added to the water how does the color of the light of the flashlight seen through the beaker compare with the color of the light scattered by the milk in the beaker? What atmospheric optical phenomena do these observations relate to?

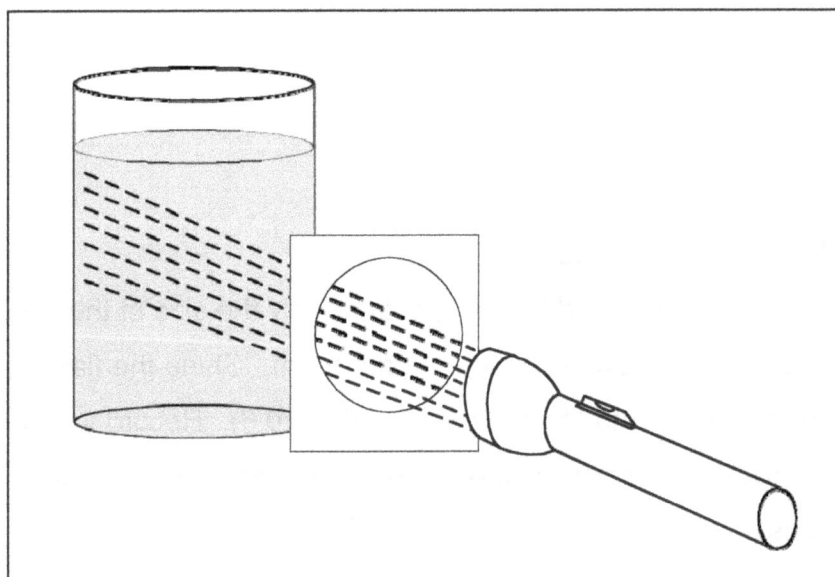

Figure 10.6 Schematic diagram

4. How, if at all, is this image different from the one you see in the sky?

5. Can a rainbow be produced from moonlight? Why or why not?

6. Why don't you see a rainbow near noon in the summer?

7. Explain why refraction adds to the length of the day.

8. If you were looking for a rainbow in the morning, which direction would you look? Why that direction?

9. Pick a sunny day and set out a sprinkler to make your own rainbow. When you see the bow where is the sun relative to the direction in which you are looking? How does the droplet size in the sprinkler spray affect the appearance of the rainbow?

Lab 52: Investigating Refraction

ACTIVITY

OBJECTIVE: The purpose of this activity is to observe the difference in refraction for air, fresh water and salt water and relate those observations to practical applications.

MATERIALS:

§ soup bowl

§ penny

§ saltwater solution (50 grams of table salt to 1 liter of fresh water)

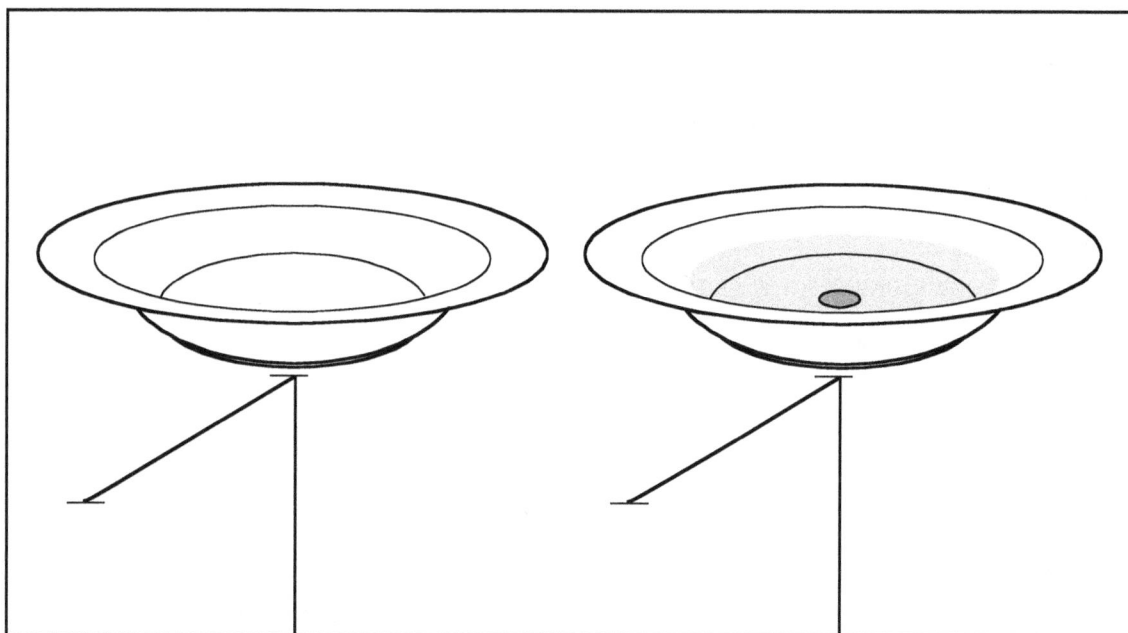

Figure 10.7 Schematic diagram

PROCEDURE:

1. Place a penny in center of a bowl (Fig. 10.7). Hold the bowl straight out from your chest at arm's length. The penny should not be visible.

2. Mark a spot on the wall or blackboard to indicate the level to which you raised the bowl. Return the bowl to this height for each procedure.

3. Add 1/4 cup of water to the bowl, making sure the penny stays in the center. Again, hold the bowl at arm's length. Do you see the penny? If not, begin adding as many tablespoons of water as necessary to make the penny visible. How many additional tablespoons did it require?

4. Repeat the procedure using salt water. How many tablespoons of salt water did it require?

QUESTIONS:

1. Which medium -- air, fresh water, or salt water -- has the highest index of refraction? Explain.

2. In this experiment, light rays reflect off the penny and travel toward the observer's eyes. Draw the light rays. (Remember: the light ray bends as it breaks the surface of the water, not when it reaches the top of the bowl.)

3. A tropical fish collector was rowing along the shore when he spotted a rare fish he wanted. The fish was swimming 8 feet below the surface of the water. Given that the angle of refraction (as the light travels from the fish to the collector's eyes) is 40 degrees, draw a ray diagram. Show the real and apparent positions of the fish.

   To calculate the apparent position, first use Snell's Law to determine the angle of incidence. Then use the given angle of refraction (40 degrees) and the angle of incidence you calculated above. Hint: You will need to use basic trigonometry to solve this problem, i.e., relationships between angles.

Lab 53: Aerosols, Visibility, and the Color of the Sky

INTRODUCTION

The color of the sky and visibility vary greatly with the time of day, the presence or absence of clouds, and particulate matter or aerosol in the air. In the absence of clouds the sky can appear red, blue, green, or even gray. These changes in color are related to refraction and scattering of light waves, which also affect visibility and exposure to burning, ultraviolet radiation. During the day a clean, clear atmosphere appears blue due to the preferential scattering of blue light by air molecules. The wavelength of blue light is similar to the diameter of air molecules making air molecules an efficient scatterer of blue light. As the sun sets, the sun's rays pass through more and more air, diminishing the blue light through scattering, until red wavelength light dominates. Aerosols add to the scattering. If the aerosol is small, such as those that result from natural turpines released into the air by forests, the result is a blue cast to the air. The Blue Mountains of North Carolina and the Blue Mountains of Australia gain their name from this phenomena.

When particles in the air are large relative to the wavelengths of visible light, they scatter all these wavelengths equally; thus cloud droplets result in white clouds (unless they are tall, in which case the bottoms appear dark) during the day. Haze droplets make the sky appear gray through scattering. Human activity (burning of fossil fuels, agricultural activity, etc.) has increased the burden of aerosol in the atmosphere during this century, resulting in a decrease in the amount of sunlight reaching the Earth's surface. Recent research has shown that areas where the aerosol input to the atmosphere is greatest have experienced cooler surface temperatures. This effect may partially offset global warming.

ACTIVITY

OBJECTIVE: The purpose of this activity is to observe the effects of scattering and absorption on light waves, using water as an analogy for air.

MATERIALS:

§ 2 to 3 clear glass containers

§ bright (Halogen) flashlight

§ toilet paper roll

§ cardboard

§ 2 to 3 pipettes or dye droppers

§ milk

§ mud

PROCEDURE:

1. To intensify the light source, make a collimator. Take the toilet paper roll and cut two pieces of cardboard into circles for end pieces. Cut a vertical slit 1/8 inch wide in each of the circular ends. Glue the ends to the toilet paper roll, aligning the vertical slits.

2. Fill the glass containers with water. Shine the light through the containers. (Note: Place the collimator between the flashlight and the glass container. It will focus the light into a more concentrated beam.) Because tap water contains little particulate matter, not much light will be scattered or absorbed.

3. Add one to two drops of milk to the container and stir well. Shine the light through the container (Fig. 10.8).

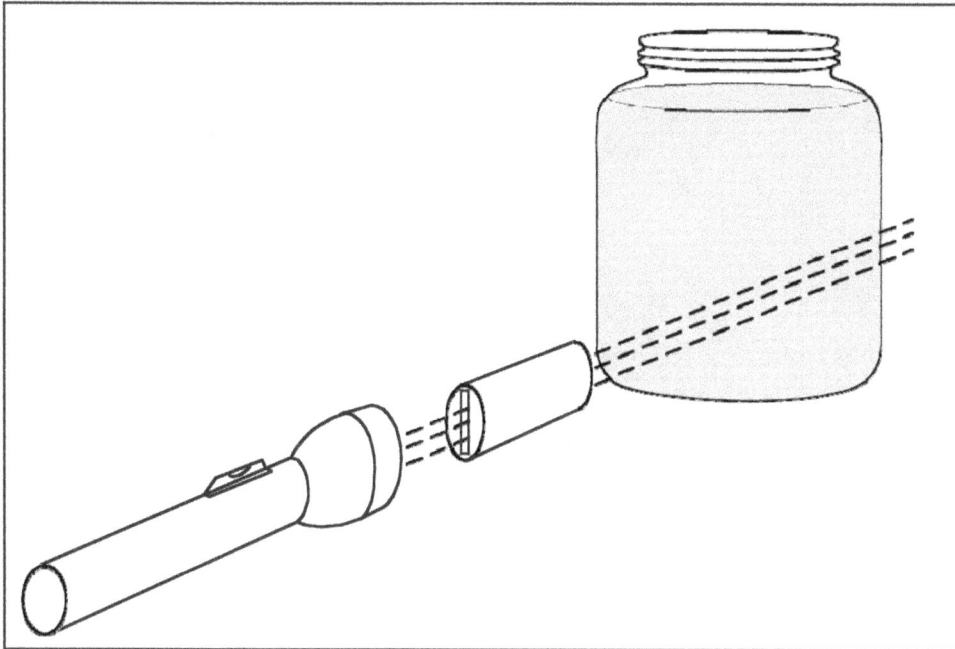

Figure 10.8 Schematic diagram

Place additional glass containers behind the first to accentuate the color.

Repeat the above procedure using mud.

QUESTIONS:

1. Look at the water in the container from the top or from a side perpendicular to the beam of light. *What color do you see?*

2. Look at the water in the container from the side opposite the flashlight. *What color do you see?*

3. What is the color of the milky water when you observe it from the top or side? Explain

4. What is the color of the milky water when you observe looking through the water toward the flashlight? Explain

5. After adding mud to the glass, what is the color of water from the top, looking down and at the sides? Explain

6. Given the statement that smaller particles scatter blue light and larger particles scatter all colors, try adding other impurities to clean water and see if you can make statements about the size of the impurities added based on your observations of the resulting colors.

Appendix: Online Resources

Links to Weather Related Servers

http://weather.hawaii.edu

http://www.wunderground.com/tropical/

http://www.nhc.noaa.gov/

http://www.nrlmry.navy.mil/tc_pages/tc_home.html

http://www.atmos.washington.edu/~ovens/loops/

http://www.prh.noaa.gov/hnl/

http://weather.uwyo.edu/upperair/sounding.html

http://www.rap.ucar.edu/weather/

http://mkwc.ifa.hawaii.edu/vmap/index.cgi

Photograph Credits

Front Cover Dust and aerosol scatter sunlight over Kaena Point Hawaii, resulting in shafts of light called crepuscular rays. Photograph by the author.

Back Cover Sun pillar caused by reflection of sunlight by ice crystals. Photo by Margi Houghton

Chapter 1 The layered structure of density, cloud water and aerosol content in the lower atmosphere is seen across the sun as it sinks into the Pacific Ocean. Photograph by the author.

Chapter 2 The explosive nature of the eruption of Mount St. Helens 18 May 1980 was due in part to high pressures created when snowmelt water was heated by lava within the mountain.   Photograph by Austin Post, United States Geological Survey.

Chapter 3 Alpine glaciers such as this one near Zermatt, Switzerland provide a record of recent climate swings. Photograph by the author.

Chapter 4 Ice crystals form as frost on a cold windowpane in Washington state. Note how the ice crystals grow at the expense of the much smaller dew drops in the left side of the image. This is the same process by which rain forms in cold clouds. Photograph by the author.

Chapter 5 Mountain wave cloud over Boulder, Colorado. As the name implies, these clouds result from air being lifted over mountains and are stationary. Photograph by the author.

Chapter 6 Heavy snow pack and rimed trees in the Washington Cascades following a series of winter storms.   Photograph by the author.

Chapter 7 First visible satellite image taken by GOES-12 on 17 August 2001.

Chapter 8 Hurricane Floyd on 14 September 1999 in an 3-D enhanced satellite photo courtesy of NASA.

Chapter 9 A warehouse fire resulting from a natural gas leak produces a thick column of atmospheric pollution over Seattle, WA.   Photograph by the author.

Chapter 10 A primary rainbow over a waterfall on the Island of Hawaii. Photograph by the author.

www.ingramcontent.com/pod-product-compliance
Lightning Source LLC
Chambersburg PA
CBHW080325270326
41927CB00014B/3103